S0-BVP-312

GLOBAL STAKES

HARPER & ROW, PUBLISHERS
NEW YORK

Cambridge
Hagerstown
Philadelphia
San Francisco

London
Mexico City
Sao Paolo
Sydney

GLOBAL STAKES

The Future of High Technology in America

JAMES BOTKIN
DAN DIMANCESCU
RAY STATA

WITH JOHN McCLELLAN

Ballinger Publishing Company
Cambridge, Massachusetts

International Standard Book Number: 0-88410-886-4

Library of Congress Catalog Card Number: 82-8747

Printed in the United States of America

Library of Congress Cataloging in Publication Data

Botkin, James W.
Global stakes.

Includes index.
1. Technology and state—United States. 2. Technical education—Government policy—United States. 3. United States—Economic policy—1971– I. Dimancescu, Dan. II. Stata, Ray. III. Title.
T23.B67 1982 338.4'7621381'0973 82–8747
ISBN 0–88410–886–4

CONTENTS

INTRODUCTION

The Future of High Technology in America

BELOW THE glittering hills of Monaco, in a Riviera hotel hugging the Mediterranean harborside, a little-noticed but highly significant meeting took place in early 1982. Founders and chief executive officers of America's booming electronics industry presented one success story after another to more than 275 of Western Europe's barons of the financial investment community. The message: Not only are America's high technology industries thriving, but they are emerging as a global economic force. Their success is leading toward a new world economy based on knowledge and advanced technology.

Within weeks, a quite different meeting of equal size and importance convened — this one in a crowded suburban hotel conference room on the East Coast of the United States. Many of the same chief executives listened to presidents of American universities and other institutions of higher education confirm the shared concern that the shortage of scientists and engineers, the lifeblood of high tech companies, would get worse before it gets better. Many universities were operating well over capacity in engineering classrooms and laboratories, faculty salaries could not keep pace with competitive industry rates, and equipment in most engineering departments was sadly outdated. And in high schools and primary schools, a dismal picture of declining standards of education was producing children who for the first time were less well-educated than their parents. Not only were reading scores down, but math and science training was diminishing as teachers left their profession for better paying jobs in industry.

America's high technology industry is a rising star in an otherwise declining economy, but whether its future will be one of success or stagnation will depend on yet another challenge: international competition. While a number of once-predominant industries like shipbuilding, steel, textiles, and now automobiles have succumbed to foreign competition, the United States still leads the world in computers, electronics, telecommunications, and other high technology fields. But continued leadership is far from assured.

International competition and Reagonomics have caught the high tech industry — and with it, the American economy — in a squeeze play. While Japan challenges our technological leadership, burgeoning defense programs soak up the engineering skills critical to continued American innovation. National economic policy is increasingly at odds with the needs of America's high technology future. In the short run, misplaced investment priorities and the high cost of capital are discouraging high-risk investments in new technological development. In the longer run, a growing shortage of scientists and engineers and our system's neglect to educate them threaten to stunt the growth of high technology companies. We seem to be spending more to sustain traditional, stagnating industries than to encourage emerging industries that are key to future economic power and social well-being.

This failure of national economic policy to focus on new industries could not have come at a worse time. Sophisticated global competition is intensifying in the high tech sector as Japanese and European companies scramble to close the technology gap. Foreign governments, especially in Japan and France, are becoming active partners in subsidizing their export industries. Their well-crafted national policies leave no room for doubt. Our competitors see that the stakes of the game are global and that the future of high technology hangs in the balance.

Much of the U.S. competitive strength in high technology happened almost inadvertently. It did not emerge from a consciously formulated national plan, as it has in Japan and France. Following World War II, America undertook a vast program to build a technically sophisticated national defense system. At the same time, the G.I. Bill provided generous subsidies for servicemen returning from the war to pursue higher education. Later, in reaction to Sputnik, America launched a buildup of technical education and resources to lead the world in the exploration of space. All these programs, along with the massive infusion of technical talent from Europe before and during the war, combined to propel America on an unforeseen and revolutionary course of technological development. From this era of government-sponsored research and development emerged the computer, semiconductor, communication, and instrumentation products that today provide the most promising foundation for new economic growth for the rest of this century. By the year 2000, for example, the high tech industry is expected to be second only to energy in its impact on the economy of America and indeed the world.

The United States, having achieved technological leadership and commercial success as an offshoot of these other objectives, has no gameplan to sustain its momentum in high technology. We take our success for granted, not appreciating the underpinnings that have supported our achievements and that will be required to maintain our leadership.

National economic policy is geared to increasing capital formation through lower taxes and less government in order to increase productivity, reduce inflation, balance trade deficits, and ultimately improve our standard of living. This policy, embedded in the new tax laws, provides accelerated depreciation of capital investments, liberalized investment tax credits, and a new twist – "safe harbor leasing" – whereby losing companies can sell their investment tax benefits to profitable ones. On the face of it, incentives for capital investment seem like a good economic idea, and in some respects they are. But a close examination of the trade-offs as to where new investment is likely to be made leads to the conclusion that this policy reinforces capital intensive "sunset" industries at the expense of knowledge-intensive "sunrise" industries. Even worse, this policy prolongs the life of dying companies and ignores the needs of growing firms. Scarce capital resources are dissipated, providing marginal returns to the national economy. Also, this policy reduces a source of government funds that could be used more wisely to repair a faltering but strategic resource critical to a knowledge-intensive era – a system of higher education on which our economic and national security will increasingly depend. In the context of tomorrow's economy it makes economic sense to invest more in education than in steel.

The needs of fast-growing, knowledge-intensive, high technology companies are significantly different from those of slower growing, capital-intensive industries. Slower growing companies generate most of the their capital from internally generated profits. Thus, accelerated depreciation and investment tax credits are important for them. By contrast, fast-growing high technology firms that are not so capital intensive look to the equity market for start-up and expansion capital. Moreover, their future growth is dependent on high risk investments in research and development (R&D). Thus, lower capital gains tax and R&D tax credits are more attractive than accelerated depreciation to high technology companies.

While the present high cost of capital discourages high risk investments in new technology, the growth and vitality of America's new industry will be impaired more by a lack of human capital than financial capital. It is significant that the cost of capital is three times greater in the United States than in Japan; what is more significant is that the Japanese graduate three times as many engineers than America on a per capita basis and also more in absolute numbers despite a population half our size. Technical talent is the raw material that feeds the growth of the high technology industry, and we have reason to be concerned about the adequacy of our supply. There is a serious shortage of engineers in America which limits the breadth and depth of product and technology development we can undertake. Underlying this shortage is an underfunded and overstretched system of education.

Neither the strategic importance of education nor its close link to high technology is widely recognized and understood in America. Characterized by rising costs and falling test scores, our educational establishment has reached a low ebb in public opinion, and governmental support is being drastically cut back. This is a "San Andreas fault" in government policy that creates a potential for disaster in Washington. A shifting economy and an unsupported system of education will create pressures throughout society. This fault in national policy is most visible in the recent phase-down of National Science Foundation support for science and engineering education at a time when demand for engineers outstrips supply by two to one. Somehow the nation has lost a strategic recognition of education that two decades ago was a national commitment.

Even more incongruous is the massive buildup of defense spending in the face of serious manpower shortages without a commitment to expand the pool of engineering resources as an integral part of the defense budget.

Will the government's demand for scarce human resources crowd out private-sector needs in much the same way that government demand for financial capital has driven the cost of capital to crippling levels? In a sense, the United States is fighting for leadership on two fronts — militarily with the Russians, and economically with the Japanese — but with the same troops, namely, our technical workforce. It is far from clear which battle is more decisive to our national security.

The incongruities in national policy, particularly with regard to the role of education, can at least in part be explained by the fact that our society has failed to grasp the full significance of the transition that is now under way. We are moving from a capital intensive, physical-resource-based economy of the first half of this century to a knowledge-intensive, human-resource-based economy in the last half. The formulas, policies, economic theories, and conventional wisdom that facilitated the earlier transition from an agrarian to an industrial society are no longer applicable to the transition now in progress from an industrial society to an information society.

It should come as no surprise that we cling to old ideas and values long past the time when evidence clearly calls for change. Research on changing patterns of history has shown that societal forces tend to reinforce the status quo and even to condemn as heretical new facts and ideas about the world and how it functions. Galileo's persecution by the authorities of his time serves as a constant reminder that our well established, well intentioned, and best-educated contemporaries can be dead wrong about what makes for a better world.

Thomas Kuhn in his seminal work, *The Structure of Scientific Revolutions,*[1] carefully analyzed and documented the dynamics of change that involve significant transitions of ideas and values. His study focused on changes in the physical sciences; later authors extended these same concepts to political, religious, social, and economic patterns of change. In the history of science, supposedly the most factual and objective of all human endeavor, Kuhn points out that established views of how the world works prevail long after a new discovery may have rendered them obsolete. For example, Einstein's theory of relativity and of quantum mechanics revolutionized Newton's mechanistic perspective of the world. Yet, at that critical juncture from old to new, nearly all the experiments, scientific work, and research policies continued to be based on the validity of Newtonian laws long after the time when these efforts could yield useful results. Today we are at a similar crossroads of economic and social history, insisting on using old theories when we already know what works better.

From another perspective, Jay Forrester, an MIT professor who is building a national economic model based on engineering feedback theory, notes that economic vitality waxes and wanes in fifty year cycles — the so-called Kondratieff long wave.[2] He explains this long-term business cycle by changes in the underlying technological base of the economy and by the tendency toward the end of a fifty-year cycle to overaccumulate capital stock in old technologies, which are subsequently rendered obsolete by developments in new technologies.

The United States is now experiencing a basic change in technological dependencies from steel and energy-based products such as cars, to knowledge-based products such as computers. Whether or not the Kondratieff wave presages further economic malaise for the mid and late 1980s, Forrester makes a significant point that there are long lags between the time when capital investments are made and when the results and payback from the investments can be measured. During periods of transition, this can lead to an "overshoot" of investment in old technology, which then produces a drag on the economy until these marginal investments are written off.

When juxtaposed with the observations of Kuhn and others that societies tend to continue their most recent behavior pattern regardless of evidence to the contrary, Forrester's theory takes on added meaning. What about the wisdom of tax policy that retains large amounts of the nation's investment resources in corporations whose technologies are obsolete and whose markets are dying? It makes one wonder why the Chrysler Corporation gets $1 billion in public financing and new high technology firms are left to compete for overpriced capital on the open market. In other words, incentives for capital investments can in fact amplify and prolong the stagflation they are intended to cure.

Robert Hayes, a professor at the Harvard Business School, also challenges today's conventional wisdom.[3] The usual reasons given for declining productivity in the United States, he says, do not hold up under close scrutiny. Such impediments as insufficient capital investment, high energy costs, labor problems, and government red tape may contribute much to our problem but little to our thinking about its solution. He observes, for example, that France and Germany have these same problems and then some; yet their rate of productivity growth has been significantly higher than ours.

Hayes concludes that meaningful strides in productivity do not come so much from incremental investments to improve existing products and manufacturing processes as from investments that create products and markets based on substantiallly new and different technologies. What is wrong, Hayes argues, is the preoccupation by American management with short term incremental investments in existing businesses to the detriment of long-term investments in new technology and market development. By this reasoning, it is technological innovation more than capital formation that will solve our productivity problems. And innovation comes from investments in people as well as investments in machinery.

From many perspectives, it can be argued that more of the same is not what is needed. The decade of the 1980s promises to be a time of fundamental change. The longer we resist and prolong the transition, the more radical and painful the adjustment will be to a new economy. To those in high tech industry, the transition has already taken place; it is a fact of life. But among the wider public and most decisionmakers, the realities of a new era have not yet been fully understood or accepted. This is why U.S. national economic policy remains only marginally relevant to the most promising segment of a new American economy and why our educational institutions are slow to respond to the needs of a knowledge-intensive society.

The fact that high technology products embody an unprecedented amount of human knowledge and technically sophisticated labor will change the equations by which national priorities are calculated. Whereas American wealth and power have traditionally been based on natural resources and on capital investment in physical plant and machinery, the balance is now tipping toward investments in people and knowledge as key resources. This is not to deny the continued importance of natural resources and the need to conserve nonrenewable ones more wisely. Nor is it to repudiate the role of capital and the need to control inflation more vigorously. But once the concept is fully grasped that knowledge should be seen as a strategic resource with an importance equal to or exceeding natural resources and physical investments, then a chain of propositions follows that will change the way national priorities and strategies are set in

America. The most important among these propositions concerns education, and the strategic long-term need to resupport and reorient the American system of education. Another concerns training, and the need to revamp our approach to retraining workers who are displaced by technological change.

The purpose of this book is to pull together in one place the facts and figures, arguments and opinions, that affect the future of high technology in America – a future course that is bound to be significantly different from the path represented by increasing defense expenditures and by reindustrializing old industries. We shall argue that present national policies are in many ways inconsistent with the experiences and needs of high technology companies, as shown by an examination of the special aspects of this growing industry and the new needs generated by these characteristics. We shall look at our competitors, especially Japan and France, and contrast their centralized approach to our nation's reinforcement of the status quo. We shall talk about deficiencies in our education systems and about the causes and cures. We shall go into some detail about competing priorities between national defense and economic development.

These challenges are producing new responses and a new mood for action in America. Especially promising are emerging models for cooperative efforts among industry, academia, and government that have the potential to solve our dilemmas. National leadership will be critical to guide us through this period of change and some ways will be suggested for company, university, and public officials to facilitate the transition. Finally, we shall present the views of important leaders in industry, academia, and government about the seriousness of the problems facing universities and high tech industries, and some of their ideas of what can and should be done about them.

We cannot hope to offer a comprehensive solution to such complex issues and problems. Our goal is rather to develop a deeper understanding of an important challenge that we face as a nation, and to get more people to think seriously about this challenge. By doing so, we hope to contribute to accelerating and easing the transition to a new and exciting era of progress and prosperity.

Chapter 1

GLOBAL STAKES

IT HAPPENED to steel, to shipbuilding, and to railroads. It happened to cars and consumer electronics. In rapid succession, a number of U.S. industries long dominant in the world market succumbed to unexpectedly aggressive international competition. Now the challenge from abroad is threatening the information technologies – computers, communications, and electronic components. Not only Japan, but also countries such as France, are systematically contesting U.S. leadership. Will Detroit's automobile legacy be shared by America's computers?

Cars and Trains: Portents of the Future?

The failure of America's automobile industry to anticipate changing markets and new production techniques offers an important lesson in the dangers of overconfident, short-sighted management. Carried away with fins and chrome during the 1950s and early 1960s, now playing catch-up in fuel efficiency and assembly line robotics, Detroit has slid into a deep, self-inflicted recession. Through smarter management, Japanese competitors have built their cars not only more cheaply – $1,750 cheaper per car on the average – but better. A recent comparison in the *Harvard Business Review* of six leading American-built cars rated against the four leading Japanese cars sold in the United States tells the story. On a rating scale of 1 to 20 with 1 being the worst and 20 the best, U.S. cars rated 9.1 on mechanical performance versus 11.7 for the Japanese. Body quality was rated 8.9 for American-built cars versus 16.4 – or almost twice as good – for Japanese cars.[1]

In another U.S. industry that once ranked first in the world, the deterioration has become legendary. The railroad industry, historically considered a strategic sector of the American economy, is technologically moribund. At present, the Amtrak passenger train from New York to Boston takes five and a half hours to complete the 230 mile run at an average speed of 42 miles per hour. A technological marvel? Maybe in 1880. But in 1981 in France, commercial train service was introduced between Paris and Lyon running at an average speed of 160 miles per hour. French advances symbolize that nation's efforts to stimulate its technological infrastructure and strategic industries toward world leadership. And for products as prosaic as subway cars, America turned in 1982 to Japan. In March of that year, New York City announced the purchase of $275 million worth of new Kawasaki cars because of superior "price, quality, date of delivery, and financing."[2]

If the U.S. lesson in cars is a failure of management in the automotive industry, the lesson in trains is a failure with national consequences. While American auto manufacturers have not adapted quickly enough to changing marketplace demands and assembly line techniques, earlier executives of the railroad industry failed to foresee the fate not only of an entire industry but one of strategic importance to many other industries. The same could be said of steel, which is also basic to many other industries and which has also been battered by international competition.

Global Competition in High Technology

The question of global competition in computers is doubly serious because, like cars, it is a large industry under attack and, like railroads, it is an intermediate industry of strategic importance to many other industries. Computers can be end products as are cars or part of the infrastructure as were railroads. In both cases challenges to the leading U.S. position are being mounted not only by the Japanese, but by other high technology minded nations such as France – the former by systematically targeting well-defined sectors and the latter in selected fields such as aerospace and telecommunications equipment. The resulting slippage in America's leading-edge technologies raises important questions about the lack of strategic focus of the U.S. economy.

A snapshot of worldwide rankings in 1980 shows little cause for alarm. Even now, the United States still holds a strong and dominant position in computers, semiconductors, and other electronic equipment in terms of production for the world market and consumption of information technologies within the domestic marketplace. U.S. companies maintain more than 80 percent of world computer market share (and Japan a seemingly modest 6 percent), and the U.S. domestic market still accounts for more than 50 percent of world sales.

THE SIZE OF ELECTRONICS MARKETS IN 1982:
(billions of $)

	U.S.	EUROPE	JAPAN
Computers & Software	$45.8	$20.2	$11.3
Semiconductors	8.2	2.6	6.3
Consumer Electronics	20.2	14.8	9.1
Other Electronics	65.5	23.2	17.0
Defense	28.3	—	—
TOTAL consumption	$139.7	$60.7	$43.7

Source: "World Market Forecast," *Electronics*, Jan 13, 1982.

But from a future U.S. perspective, the outlook is not so reassuring. While an erosion of America's world market share is to be expected as both the industry and competing economies mature, the speed of the U.S. decline in certain sectors is alarming to many industry leaders. An example is the 16K dynamic RAM chips battle.* In only a few years, Japanese manufacturers captured an astounding 40 percent share of the world market for these devices. The next round of the fight is for 64K chips, and the victor could stand to gain an estimated $2 billion by 1985. Japan has taken a surprising early lead, capturing 70 percent of the initial market.

The 256K chip market will be next. Japanese aggressiveness and boldness in developing new market opportunities catches us off guard. A case in point is Nippon Electric Company's plan to build, at a cost of $100 million, one of the world's most advanced semiconductor plants. Located in California's own "Silicon Valley" and employing more robots and fewer workers than any comparable facility, it is scheduled to produce a

* See *Counting K's: 16, 64, 256* . . . at end of chapter for explanation of technical terms.

new generation of 256K memory chips when it goes into operation in 1983 or 1984.

Another case that gives American executives pause for thought is Fujitsu's announcement to compete head-on with IBM, its giant U.S. rival in the global computer market. As a first step, Fujitsu, aided by protectionist policies, has replaced IBM as the top seller of computers in Japan, a computer market second in size only to that of the United States. And the announcement by the Japanese government that it has launched a ten-year program to create the fifth generation computer opens the door to a truly global economic contest for new markets.

To attribute similar competitive threats to other nations such as France comes as a surprise to most Americans. Few are ready to believe that France is a market challenger of U.S. computer and telecommunications hardware and software technology. A study completed in 1980 showed that American software vendors perceived France to rank a distant fourth in annual sales behind the United States, England, and Germany.[3] In fact, by 1980 France's $1.5 billion a year software and computer services industry was already second to that of the United States.[4] Moreover, French hardware manufacturers such as Matra, Thompson-CSF, and Alcatel Electronique are becoming increasingly familiar names among industrial leaders as France increases its efforts to capture market share not only in the United States but throughout the world.

It is no comfort for American managers to consider France's election of a Socialist government in 1981 as a potential diversion from aggressive international competition. The François Mitterrand government was quick to establish the continuity of the technological growth strategies inherited from Valery Giscard d'Estaing. The new thrust was underscored by a large increase in the national research and development budget. A few months after his election, Mitterrand outlined the broad framework of his government's high technology program for electronics and communications. Speaking to industry leaders, he said, "I want to set before you a challenge: to establish France, by 1988, as the world leader in as many sectors of the communications industry as possible."[5]

Mitterrand's resolve is given added weight when one considers his nationalization efforts, put into effect early in 1982. They have brought one half of the French informatics industry under public control, affecting about 38,000 employees and $2.5 billion in sales as of 1980.[6] Behind this new public agglomeration is the financing power of the French treasury with a promised $5 billion investment scheduled over five years.

Another clue to future market competitiveness of the United States, Japan, and Europe can be found in the educational strategies of these nations. A useful weather vane is the number of young people graduating as electrical engineers. Japan graduates more electrical engineers than any of the other countries shown below. At the same time, computers and electronics account for a relatively greater share of the Japanese national economy than is the case in other countries.

THE ECONOMIC IMPACT OF ENGINEERS: Who's Leading?

	Electrical Engineering, 1977		Computers & Electronics, 1975	
Country	Number of Graduates	Degrees Per 25,000 Pop.	as percent of GDP	$ value of shipments*
Japan	21,090	4.6	4.3%	$20.9
United States	14,085	1.6	2.7%	39.8
West Germany	6,649	2.7	2.1%	8.9
France	1,749	0.8	2.3%	7.7

Sources: Adapted from International Monetary Fund, *International Financial Statistics Yearbook, 1981*; Semiconductor Industry Association, *The International Challenge*, May, 1981; and "A Report by the Sector Task Force on the Canadian Electronics Industry," *Planning Now for an Information Society*, Science Council of Canada Report #33, Ottawa, March, 1982.

* Figures in billions of dollars.

No U.S. National Policy

Raising the question of the need for national policy may seem paradoxical for a nation that is already parent to IBM with $29 billion in annual sales and nearly 50 percent of the world mainframe computer market. Yet the success of the Japanese and French, both of whom have explicit national strategies, as well as the increasing importance of computers, electronics, and telecommunications to the United States and the world economy, is cause for soul searching. Is some form of national high technology growth policy needed, and if so, what form should it take?

The question of national policy is being raised with increasing frequency. Peter Drucker sums up the post-war successes of the Japanese: "They based themselves on the novel premise that national economic policy begins with a careful assessment of the world economy. . . Japan tried to manage supply to fit the demands of the world economy; they succeeded. . . The United States tried to manage demand to fit domestic political goals; they failed."[7] MIT economist Lester Thurow puts it more directly, "Americans simply will not be able to compete in the modern world of international trade without changes in the way they traditionally operated."[8]

For Americans, discussion about national strategy does not come easily because an implied centralization of federal power is perceived by some as antithetical to the culture. But equally discomforting is the sense of slipping behind. Again Thurow:

> When you combine an inevitable loss of technological and economic superiority with a brave new world where Americans are for the first time dependent upon international trade, it is not surprising that there is a feeling of unease. It is uncomfortable to know that there are others who are smarter, harder working, and better organized who may permanently eclipse you.[9]

Any concept of national policy in the United States is likely to be vastly different from those in Japan or France. Nor is there any assurance that countries with successful strategies in place today will remain at the leading edge of the economic and social transition tomorrow. Nonetheless, the U.S. slippage in the global economy, whether temporary or long term, suggests that we seriously reexamine the role that national policy could play in the U.S. context.

To regain the initiative in the 1980s and 1990s, American policymakers need to shift their thinking from a domestic economy based on abundant resources to a global economy with scarce knowledge-based resources. In addition to the conventional financial and tax considerations, an important part of any national policy will be the role of education and the quality of the work force in high technology jobs. At least part of the Japanese success, and to some extent that of the French, has been to bridge this conceptual chasm from domestic to global and to make human resources an inherent part of the strategy.

COUNTING K's: 16, 64, 256 . . .

Technically, the 16K RAM (random access memory) is a huge step up the memory ladder first fashioned a decade ago by Intel when it pioneered the 1K dynamic RAM. The K stands for 1,000 memory cells (actually 1,024, to be exact), while dynamic means that the chip must receive continuous tiny electric pulses or lose its memory, much as a human brain must have oxygen or die. Random access means that each memory cell can be reached as though it were a telephone waiting to be dialed or rung. The 16K chip, 16,000 memory cells concentrated onto a single chip, is now a basic component in many computer-related devices.

Engineers have recently developed the 64K dynamic RAM chip by compressing thirty or more memory cells into a space no wider than a human hair. Since their introduction in 1978 when Fujitsu announced product breakthroughs, the scene crowded rapidly with competitors, led by both Texas Instruments and Motorola. But to the dismay of American producers, Japan quickly captured 70 percent of the world market. Only two major U.S. producers are presently in the running — Motorola and Texas Instruments. Hitachi has 40 percent of the world market, Fujitsu 20 percent, and Nippon Electric 6 percent, while of the U.S. companies Motorola holds a 20 percent share and Texas Instruments has 7 percent.

The next dynamic RAM, already under development and again with strong evidence that Japan is ahead of the pack, will be the 256K chip. It won't be long before we reach the megachip, but which companies will reach megabuck sales?

Source: Adapted from "Japan's Ominous Chip Victory," *Fortune*, December 14, 1981, and from "The Japanese Assault," *Industry Week*, April 20, 1981.

Chapter 2

THE NEW ECONOMY

THE EARLY 1970s were a milestone in economic history. It was the period when two converging paths merged – the calculator met the computer and a new breed of low-cost microcomputers was born. This marriage marks the coming of age of a growth process underway since the 1940s – a basic change in America from an industrial to an information economy.

Microprocessors, memories, and other integrated circuit chips are becoming the critical raw material to the information economy just as oil became the predominant fuel in the industrial economy. This means not only a rapidly growing electronics industry but, along with it, a new lease on life for older industries as innovation in the semiconductor technologies that form the basis of microcomputers accelerates and spreads throughout American industry. The net result is that a structural shift is under way in the U.S. economy. High technology industries have become a major force whose new and sometimes explosive growth is having an impact on almost every other industry and function of society. This impact, with its worldwide repercussions, is creating a new economic picture.

Old Thinking, New Realities

When Ronald Reagan was elected president in a period of economic decline, the call went out for "reindustrialization." The result has been a program of tax cuts to stimulate reinvestment and to rekindle productivity primarily in the older and now-matured industrial sector. An important question in many people's minds is whether these policies will work. A more haunting question is whether *any* policy geared toward old industry can work.

Contemporary economic difficulties faced by the United States and its industrial trading partners have only partly to do with industrialization or reindustrialization. Rather, the most pressing issue concerns a fundamental shift in the world economy, which drastically changes the rules governing economic and social success.

Most Americans still equate the nurturing of new economic power and growth with industries based on a prior generation of manufacturing technology, like cars and General Motors, or on traditional exploitation of raw materials, like oil and Exxon. This is mirrored in most senior corporate management meetings. Discussions still focus primarily on investments in physical plant and machinery rather than, for example, on investments in human resources. It is easier to allocate money for renovating a building than for upgrading an employee's capabilities. Tangible assets are more valued than intangible ones. Products still take priority over people. These are truisms of an old industrial order.

The economic problems of America's old industries are real enough. But our chief shortcoming is our persistence in applying old thinking to new situations. Many fail to comprehend that a long-term cycle of old industry is being transformed or even superceded by a new period of knowledge-intensive economic activity. And many others, while they recognize that a new era has begun, fail to act accordingly.

The shift can be described in many ways: from products to services, from physical resources to human resources, from investment in machinery to investment in knowledge, from capital intensity to knowledge intensity, from a domestic economy to a global economy. Fundamental to a successful transition is the foresight to understand and nurture the health of future-oriented industries – as the Japanese call it, a move from sunset to sunrise industries.

For several decades, the United States has led the world in launching the new economy. Yet in many ways the national will to surge forward and sustain that lead has stalled, leaving Japan and others as apparent heirs to abdicated U.S. leadership. At present, the U.S. economy seems poised near the midpoint of a historical transition. About half of the American gross national product (GNP) comes from manufacturing and farming, the other half from "information jobs." A significant and growing fraction of these information jobs, about 15 percent in 1975, are directly associated with computer and electronic technology.[1] The future of this sector will have a powerful impact on the economy as a whole.

The United States in Transition

Since the early 1960s, when Harvard sociologist Daniel Bell coined the term "post-industrial society," much of the academic world and a significant part of the business community have been aware that the economic underpinnings of American society were changing. Just how they were changing was described by Fritz Machlup in a series of books entitled *Knowledge: Its Creation, Distribution, and Economic Significance.*[2] More importantly, how much they had changed was measured by Marc Porat in his classic work, *The Information Economy.*[3]

Porat documented how in 1967 the largest sector of the U.S. labor force (46 percent) was engaged in information jobs. He also showed the startling fact that industrial workers had gone from a peak of 65 percent of the labor force in 1950 to an estimated 25 percent today. But Porat's lasting contribution was to demonstrate that the concepts of "knowledge, communication, and information," as outlined by his mentor, Machlup, could be identified and measured. The notion that knowledge industries, knowledge occupations, and knowledge workers had a quantifiable basis in economic reality was novel. It was also a powerful idea that inspired not only a vanguard of American thinkers but also the mainstream of Japanese policymakers.

SIGNS OF A SHIFTING ECONOMY

U.S. Workforce in Information Jobs

1880	5%
1920	15%
1940	25%
1960	40%
1980	(est.) 48%

The U.S. Employment Profile, 1979:

services & information	72%
manufacturing	25%
agriculture	3%

The Japanese Employment Profile, 1979:

services & information	54%
manufacturing	35%
agriculture	11%

Source: American figures, *The Information Economy*, M. Porat, 1977
Japanese figures, *Japanese Yearbook*, Tokyo, 1979

Knowledge-Intensive Industry

What is knowledge-intensive industry? Basically, it is an industry of companies whose primary activity is production, manipulation, or distribution of information goods and services. Some activities such as telecommunications or data processing are immediately obvious. Others are more difficult to classify. Is a management consulting company, which surely produces knowledge, still to be considered knowledge intensive if its advice turns out to be wrong? Universities are knowledge intensive, but are they industries? What about companies that produce not information itself but equipment such as word or data processors that manipulate information?

These questions touch so many bases that answers will never be definitive. Machlup discovered fifteen different definitions of "knowledge" in the English language alone, and no further clarification comes from the fact that most other languages have at least two words for the verb "to know" (French "connaitre" and "savoir"; German "kennen" and "wissen"). The best definition is by relative comparison.

Many contemporary writers, from the English historian Arnold Toynbee in *Mankind and Mother Earth*[4] to American futurist Alvin Toffler in *The Third Wave*,[5] compare and differentiate three historical ages: an agricultural age, an industrial age, and an information or knowledge age. Establishing a line to delineate the ages is not easy. Equally difficult though solvable is the problem of specifying to which period a particular activity or industry most belongs. Would a tractor be classified as part of the agricultural or industrial age? Would a dietician be considered an agricultural or information worker?[6]

Classification difficulties notwithstanding, these three distinctions are useful. "Knowledge" and "information" can be contrasted to "agricultural" and "industrial" to describe predominant economic and social characteristics of an era. In a strict sense, knowledge is not synonymous with information. For the purposes of this book, we use a tighter specification by adding another word pair: high technology. "High tech," as the term is often used in shorthand, refers to the application of science to products that are at the state of the art in terms of their function and design. Note that high technology connotes these two attributes — it is *applied* science as well as *state-of-the-art* knowledge.

Examples of knowledge-intensive, high technology industries are computer manufacturers (and all of the associated electronics and communications activities), the new biogenetics companies, pharmaceutical firms (which are also engaged in biogenetics), the chemical industry, and aerospace companies. All of these enterprises are inordinately dependent on knowledge, and make extensive use of science and technology.

However, not all segments of these industries that use science and technology extensively are at the state of the art. Some are no longer changing dynamically or growing rapidly. For example, the electrical power industry was once a pace setter but has now stabilized in terms of the equipment and processes it uses. Also, the manufacture of many chemicals is now routine and no longer considered high technology. Today, the state of the art in chemicals comes from genetics. Genetic engineering is to the chemical industry as computer sciences are to the electrical industry.

To make our point more clearly, we will concentrate exclusively in this book on those knowledge-intensive, high technology industries that the French call "informatique" and that encompass computers, microelectronic components, instruments, telecommunications, and other information technologies, software, and services.[7] But the conclusions reached are applicable to other emerging science-based industries, such as biogenetics; the new materials sciences, such as ceramics; parts of telecommunications, like fiber optics; and aerospace, aeronautics, and other space-related ventures. These relatively new businesses are considerably different from older, now-matured manufacturing enterprises even though these too are feeling the effects of the information revolution.

The Role of Capital and the Importance of Exports

New business can be differentiated from old ones in many ways. One way is by comparing capital requirements. While all businesses require capital for investment in equipment, buildings, machinery, and other material goods, the new high technology companies need proportionally less than many others. The petroleum industry, for example, invested $108,300 per employee in 1975 versus under $20,000 for electrical machinery or instrumentation. Transportation, tobacco, chemicals, and paper are other industries that require a high proportion of capital investment per employee.[8] Investment in new technologies can be highly productive, but due to rapid technological change, investments in capital goods is less productive for many companies than investments in human resources, which can be self-renewing.

For high technology companies, the access to human resources has an importance that equals and even exceeds the availability of capital resources. In "Trends in U.S. Technology,"[9] Michael Boretsky shows that the new knowledge-intensive industries are characterized by a workforce that includes scientists and engineers – in non-R&D functions such as sales, marketing, and manufacturing – in numbers five times that of mature, low technology industries. The labor force requires 70 percent more skilled workers than traditional manufacturing industries. And R&D expenditures run, on the average, 10 to 12 times that of nontechnologically-intensive products. Altogether, about one-third of the workforce hold college degrees, of which more than one-half are technical degrees. Another one-third hold associate degrees requiring two or more years of education beyond high school.

In their book *Minding America's Business*, Magaziner and Reich point out the relationship between capital investments and exports.[10] Between 1970 and 1979, the cumulative capital investments of the auto, steel, and paper industries totalled $75.4 billion. Yet all three showed significant negative declines in their net export values.[11] In contrast, the performance of the computer, materials handling, and aircraft business show a capital investment of only $18 billion for the same period, but a positive net export gain for all three of $9 billion. This suggests that while U.S. investment rates have probably remained as high as those of other nations, our investments went to low-growth, old industries at rates four to five times that of high-growth sectors such as computers and aircraft.

So far, America has been preoccupied with a defensive posture of protecting its internal market from erosion by imports. But to maintain and enhance its standard of living, America must take greater initiatives in its international strategy, making sure it has something of value to exchange for ever-increasing imports. Since the early 1970s, the United States has begun to evolve a unique international export strategy. Two of our most successful exports are agricultural products and high technology equipment. The fact that our largest imports are petroleum supplies puts us in the position of selling computers and food to pay for oil.

Not only have exports become critical to the U.S. economy but knowledge-intensive, high tech industries are critical to the success of an American export strategy. In 1980, two of the top twenty U.S. exporting firms were selling food and forestry products abroad, most notably grain to the Soviet Union. Of the remaining eighteen, four were computer-related firms, which represent the fastest growing sector of our export economy. Companies in the information technology sector usually export an above-average proportion of their production. Typically between 25 and 45 percent of their sales are outside the United States. In 1980, this translated into a $6 billion contribution to the U.S. balance of payments. If all R&D intensive products and services are counted, the U.S. balance of trade surplus has tripled between 1967 and 1977, from $8.8 billion to $27.6 billion.[12]

This positive performance in international trade is becoming increasingly important to the United States as an ever larger percentage of the GNP is based on materials and goods that must be imported from abroad. The 1960 GNP figures showed only 10 percent of the economy accounted for in exports and imports; by 1980 the proportion had jumped to 25 percent. American insulation from the world economy has given way to interdependence, which makes exporting of goods and services essential.

Another striking and significant difference between old and new industry is the different emphasis on research and development (R&D). At a time when industrial R&D spending for the American economy as a whole has slumped, falling by 25 percent since 1965,[13] it has shot upward for high tech companies. Of the top ten U.S. R&D companies, all are in the information technology industry in terms of the percentage of sales allocated to R&D activities.[14]

The transition under way in the U.S. economy forces us to rethink other long-held assumptions as well. John Naisbitt, former senior vice president of the trends-forecasting firm Yankelovich, Skelly & White, questioned why the recessions repeatedly predicted in 1979 and 1980 never seemed to materialize fully.[15] The reason stems from a failure to understand that the United States is a dual economy suffering an industrial recession but enjoying an information boom. Thus one part of the economy can be in recession while another is not.

The role of labor unions is similarly differentiated. While labor unions are declining overall (from 32 percent of the work force in 1950 to 19 percent in 1980), the significant fact is that they are becoming proportionally stronger in the shrinking industrial economy while building relatively little power in the growing information economy. Knowledge workers rely less on organized labor than do production-line workers in traditional industrial society.

While the impacts on recession and unionization are important, the more critical impact is the one on strategic resources. In an industrial society, the critical resource is capital invested in physical plant, machinery, and technologies that multiply the muscle power of labor. As society shifts toward knowledge intensity, the critical resource becomes not conventional capital investments in machinery, but the investment in people.

What does "investment in people" mean? Most typically, it means money for education, training, and research. It also includes investment in physical equipment such as word processors or computer systems to support people's activities or, even more so, the software that operates the equipment. It is not unusual to find a company investing four times more on software development than on the machine that runs the software. But the real meaning of investment in people encompasses more than this. It is the codeword for a new economic outlook where human resources, with their stock of information, education, and knowledge, become the key resource and where human ingenuity, to learn, innovate, and communicate, becomes the key to increasing productivity and society's standard of living. Also, it is retraining and reeducation to keep our work force abreast of changing technology, and to accommodate transitions from sunset to sunrise industries. It is this premium on investment in knowledge that lies behind the term "knowledge-intensive" industry.

The Global Dimension

Unlike agriculture and manufacturing, the new knowledge-intensive industries were born global, and their direction and fate will be determined more by international developments than by any single national policy. The ideas that shape high technology move quickly across borders – most quickly within the advanced market economies, but also to and from Third World countries. Brazil, for example, is developing its own computer industry (1981 gross sales were $1.7 billion). This universal quality of the new economy explains how organizations like IBM can be world companies as well as American, and why company or country policies need to be global in scope as well as nationally based.

WHERE ARE THE COMPUTERS?

	1960	1970	1978	1983[a]	1988
United States	5,500	65,000	200,000	400,000	700,000
West Europe	1,500	21,000	110,000	225,000	450,000
Japan	4 00	6,000	45,000	70,000	140,000
Others	1,600	18,000	95,000	205,000	460,000
TOTAL	9,000	110,000	450,000	900,000	1,750,000

[a]Figures for 1983 and 1988 are forecasts.

Source: Diebold Europe (1979).

Among the leading international competitors, the United States and Japan are the most advanced, but other countries, especially those in Western Europe, also play significant roles.[16] England, Germany, Scandinavia, Holland, Italy, Switzerland, Austria, and France all have burgeoning electronics industries. Of these, France is the most interesting for two reasons. First, it presently has the fastest growing economy in the West (just behind Japan's growth rate, which leads the developed countries). Second, its position is a mirror image of the American one. Whereas the American economy has changed before its thinking has, in France the thinking has changed before the economy has. If the French think without acting, and Americans act without thinking, who's the beneficiary? Probably the Japanese, who incorporate American action and French thought to make successful strategy.

In 1978, the publication of a French "white paper," commissioned by former President Giscard d'Estaing, took the country by storm. *L'Informatisation de la Société (The Computerization of Society)*,[17] written by Simon Nora and Alain Minc, showed two things: that internationally, France's economic survival depended on successfully exporting computer-related services and technologies; and that domestically, French social policy would be shaped in large measure by the performance of the export strategy. Among much of the French readership, the major issue was social policy and whether the international strategy would lead to greater centralization or decentralization of society. In the United States, where the report is barely known, we need to focus our attention on national strategy for a new world economy as it affects American living standards and social policy.

The transition to knowledge intensity is not an isolated trend but occurs at a time of widespread turbulence. At least four other trends are presently conspiring to change the economic and social life of developed countries: the shift from abundance to relative scarcity, the dilemma of rising expectations and falling productivity, the emergence of the North-South dialogue, and the increasing social demands for further decentralization and more participation. Which shape a knowledge-intensive society will take over the next few decades depends largely on how it relates, or fails to relate, to these other trends.

A New Engine for New Growth

While recession may temporarily slow the growth of high technology, the expansion of the information technology sector in the world economy is expected to be very large. During a five year span, sales of small, portable computers in the United States are estimated to grow sevenfold – from $945 million in 1980 to $6.5 billion in 1985. Software sales are expected to jump 400 percent, from $1.7 billion to $7.5 billion.[18] New computer companies are sprouting daily. In a few short years Apple took on a new meaning and became a household word. And now Osborne in little more than a year, is giving Apple a run for its money.

The marketplace for electronic goods, from chips and microcomputers to optical cables and electronic battlefield equipment, totaled to $105 billion in 1980; two years later it was $140 billion; in 1985 it should reach $205 billion, or a value equivalent to 30,000,000 cars. The information technology industries are already larger than the auto industry, and they are envisioned to be second in size only to the energy industry within twenty years.

High tech industry's growth in sales is topped only by its explosion in technological advance. A number symbolizing the speed of technical progress is 10^6, or 1,000,000. Computers have advanced by a factor of one million since their introduction. This means that an electronic function that can be accomplished for 2 cents today would have cost $200,000 in 1950. Or, in terms of information storage, an entire encyclopedia can be stored on a surface equivalent to the size of an old IBM punched card. By the mid-1980s, storage devices the size of a human head will be capable of holding about 10^8 bits of information. Since the brain is estimated to hold about 10^{14} bits, this has led to the projection that computer memory will surpass the brain's storage capacity in the early twenty-first century.[19]

A comparison of the pioneering ENIAC computer, vintage 1946, with the Fairchild F8 Microprocessor of thirty years later is a vivid measure of the technological progress sweeping through the electronics industry. The F8 is 300,000 times smaller than the ENIAC, consumes 56,000 times less power, is 60,000 times lighter, and is 10,000 times more reliable (the average time between failures is measured in years instead of hours).[20]

For a world concerned about resource scarcities and environmental protection, chips represent a source of lean, clean growth. They are energy efficient, resource sparing, and environmentally sound. If the automobile industry had followed the path of computer manufacturers, today's Rolls Royce would not only cost just $2.50 but would get half a million miles per gallon.

What about jobs? While the new engine of growth accelerates, does it provide enough new employment to offset the loss of jobs to robots on the assembly line or to automated work stations in the office? If economic growth outruns productivity increases, the net result is employment. Without growth, jobs are lost. A real race is shaping up between economic growth and productivity gains, between inventing new jobs and phasing out old ones.

In the United States and Japan, high tech industries are a major source of new jobs. The Bureau of Labor Statistics projects that between 1979 and 1990, the fastest growing employment sectors will be office equipment, computers and peripheral equipment, and medical services. In Massachusetts, employment in high tech companies has nearly doubled between 1958 and 1978, while it has declined 5 percent in other industries and is down 44 percent in textiles, apparel, and leather over the same period.[21] Even if this employment growth should abate in the hardware sector – which it may do as chip production techniques themselves become automated and laborsaving – the software sector has a seemingly endless appetite for new employees. A recent Japanese study showed the need for 80,000 software engineers in 1975 expanding to as many as 796,000 by 1985.[22]

In Europe, however, many employment projections for the overall economy are pessimistic. A recent German study shows computerization leading to 25 percent fewer office jobs in the private sector and up to 38 percent fewer in public administration by 1990. The German electronics industry is too small, representing only .9 percent of the labor force, for its growth to offset such projected losses. Germans are in the unfortunate position of importing unemployment every time they import microchips.[23]

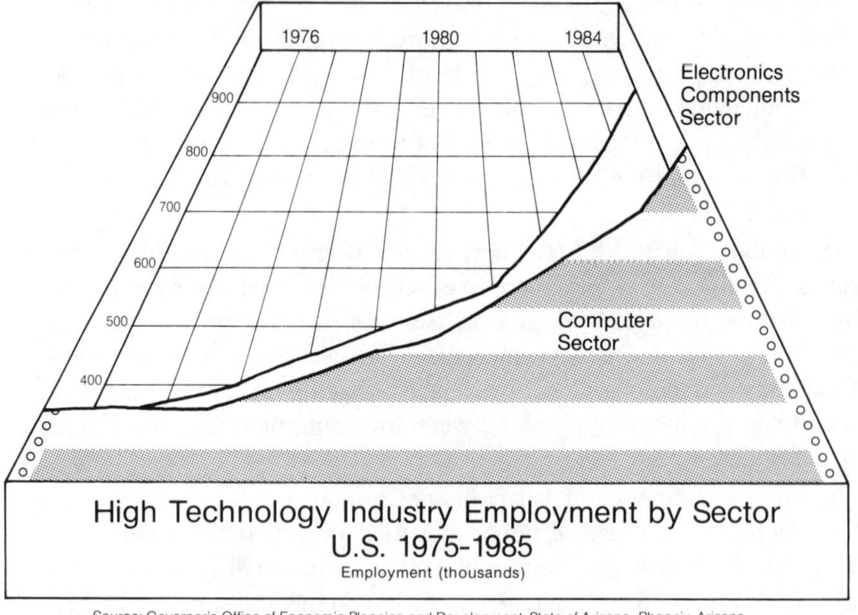

High Technology Industry Employment by Sector
U.S. 1975-1985
Employment (thousands)

Source: Governor's Office of Economic Planning and Development, State of Arizona Phoenix, Arizona

These facts contain two important strategic lessons. The first is that a strong domestic electronics industry can provide jobs at home if it stays ahead of international competition. The second is that to use the new job opportunities effectively, old workers need retraining and the coming generation of workers will need new types of education.

There is little doubt that in terms of environmentally safe growth, technological innovation, and new job creation, the information industries represent a key strategic resource to the future economic well-being of the United States. Other countries, notably Japan, have recognized this as well. For the 1980s, the Japanese Ministry of International Trade and Industry (MITI) envisions a revamped Japanese industrial base and a society more highly dependent on a knowledge-intensive, high technology, resource-efficient economy. Technology as the key to economic security is at the center of the strategy. In this sense, Japan's emphasis on the electronics and information industries is based on an economic perspective more comprehensive than a narrower industry-by-industry policy. It constitutes a national plan for the future of Japanese society.[24]

An important message of this book is that a proper perception and management of the transition to an information economy will provide positive answers to many troubling economic and social questions. Inadequately understood and poorly managed, the transition will lead to economic decline and to exacerbated social tensions. The actual outcome, however, depends on another strategic resource – higher education. Presently declining, higher education is an enterprise of rising importance. In terms of old industrial thinking, education plays a relatively minor role. But in terms of the shift toward knowledge intensity, education takes on strategic importance.

Strategic Importance of Education

Boretsky's workforce profile cited earlier gives an indication of the role that higher education is expected to play. With 10 to 12 times more R&D, 5 times more scientists and engineers in non-R&D functions, and 70 percent more skilled workers, knowledge-intensive industries are inherently connected to education.[25] For the first time in history, the link between higher education and high technology has become direct – so direct that when higher education falters, high tech can fall. In the absence of productivity increases, the output of knowledge-intensive companies could decline in direct proportion to decreases in the number of graduates from higher education and, to use Boretsky's terms, at a rate of decline anywhere from 5 to 12 times faster than in traditional manufacturing industries.

A report done by the Data General Corporation warns, "there looms on the not too distant horizon a massive shortage of human resources . . . that will dwarf all other obstacles to our company's growth and could topple the U.S. high technology and computer industry."[26] A detailed picture of the evidence of human resource shortages, drawn from studies by the National Academy of Engineering, the American Electronics Association, and others, will be presented in Chapter 4. What it will show is not only that a serious engineering shortage exists now but that it will grow more serious with time. A capacity constraint is evident in the nation's engineering colleges and universities, and they are in no position themselves to alleviate it without considerable help from industry and government.

Not everyone agrees that there is a significant long-term human resources shortage.[27] To the extent a temporary imbalance exists, they say, market forces should be left to correct it. Entering students, aware of lucrative job offers in high tech industries, will vote with their feet and apply their minds to education in the relevant fields of science and engineering. Public and private school officials, aware of the new demands by new students, will reallocate the resources needed to supply faculty and equipment necessary for a scientific education. But even if this mechanism does work, it requires time.

In an earlier period of history, America might have been able to wait for the educational system to readjust. But at least one factor in the present situation makes such a wait costly – global competition. Japan has already switched its priorities to a knowledge-intensive society and by 1980 was competing head-on with the United States, not only in automobiles and televisions, but in the computer and electronics industry itself.

Some European countries, although not to the extent of Japan, are also mounting a competitive challenge to the United States. France is vying with the United States to capture markets in telecommunications partly in the American market and especially in Third World countries in Africa, Asia, and Latin America. But the significant fact of most importance to American high tech industry and the U.S. economy is that global competition has become a reality. It is this international challenge that conditions human resource policies, high technology strategies, and associated higher education philosophies and funding. It is this challenge by global competitors to which we now turn our attention.

Chapter 3

JAPAN AND FRANCE

FOR MANY nations the post-war challenge was clear: modernize domestic economies; grasp opportunities in a widening global marketplace; and encourage national policies that would bring economic parity with the United States. Both Japan and France followed separate paths in meeting those goals. Their experiences, while culturally and politically distinct from any model the United States might envisage, provide useful measuring sticks for America's own future development strategies.

Norihiko Maeda, director of the electronics policy division of the Ministry of International Trade and Industry (MITI), recounts Japan's economic prospects as seen in the early 1950s.

> Competitive advantage theory suggested a rational choice of promoting labor intensive industries. This would have led us into a common model of Asiatic stagnation. Instead we opted for a decision to promote capital intensive and knowledge-intensive industries . . . We won that gamble. If we look at the primary exports of Japan, one notes historically that silk gave way to cotton textiles, then to ships, then to iron and steel, and more recently to automobiles. It is in this context that our strategy was established for the computer industry.[1]

Early on it became evident to Japanese leaders and planners that the post-World War II era would usher in unprecedented economic competition on a global scale. What was less evident at the time was that the competition would center not on the traditional economic factors such as cheap labor (of which Japan had an abundant supply) or on raw materials (of which from oil to ores Japan had none) but rather on knowledge. Thus it was a bold move, and one initially resisted by many Japanese planners, that led the nation to embark on a course toward knowledge intensity that is now, thirty years later, outpacing the rest of the industrialized world.

The history of French strategy starts later. While Japan was in the early stages of formulating new economic policies, France was still fighting costly wars it would not win. In 1954 its armies were decimated at Dien Bien Phu. What remained of French military strength was drained further until 1961 in a losing battle to hold Algeria.

As France emerged from her costly colonial disengagement, a new national preoccupation marked the closing years of the Charles DeGaulle decade (1959-1969). Economic and political autonomy were central to his ambitions. The effect they would have on electronics and computer policy was described in 1967 by Michel Debré, then Minister of Finance,: "Our aim is to create a solid data processing industry that will stand on its own in French and foreign markets, to the end that our nation remains master of its own destiny."[2]

These intentions were explicitly laid down in a series of special plans known as the "Plans Calculs" which were designed to put France at the forefront in "informatique" and "télématique" – the information technologies, telecommunications, component parts, software, and the associated infrastructure. The first Plan Calcul, begun in 1966 and lasting a decade, was accompanied by the founding of the Compagnie Internationale d'Informatique with $88 million (1967 dollars) to finance mergers among computer-related companies. The three Plans Calculs periods, running through 1980, stimulated the growth of France's mainframe computer business. The components infrastructure was the focus of a new plan initiated in 1978 under the title "Plan Circuits Intégrés", with five companies targeted for co-venture relationships with U.S. firms. In early 1982, the government revived its ambitions to promote the minicomputer sector with the announcement of a five-year $100 million "Plan Mini." From these initiatives emerged a two-way working relationship between government and industry, crystallized by the sweetner of government subsidies and contracts.

While Japan and later France developed explicit national strategies that were global in scope, the United States continued to rely on domestic competition at home and multinational corporations abroad as cornerstones of its economic policy. No attempt was made to coordinate the actions of individual industries or to coordinate domestic social and economic policy with multinational activity. This often led to difficulties. U.S. companies were subjected to antitrust and other regulatory policies at home while trying to compete with foreign companies whose governments were encouraging concentration by providing not regulation but protection. Whether the United States could have done differently – at a time when national attention was focused first on civil rights and then on Vietnam – is just as problematic as whether such a strategy would have been desirable or successful.

Whatever the reasons why the United States did not develop a national strategy, the future calls for a reappraisal. Should the nation have a policy, or at least a shared consensus, for moving toward knowledge intensity? Who could or should take the lead in advancing any strategy, and what should be its focus? Should a national education and research policy play a central or peripheral role?

It is unlikely that Japanese or French strategies, as they exist today, will provide a blueprint for the United States. The histories, values, needs, and opportunities of the three countries are too diverse for simple imitation. But general principles, or lessons, can be learned from the examples of Japan and France.

Japanese Strategy: "Nemawashi" and Three Principles

The basis for Japan's economic successes is a subject of intense curiosity, both casual and scholarly, to many experts, specialists, journalists, and pundits. It is commonly recognized that a variety of already well-worn explanations seek to elucidate those successes. Among them are such tactics as restrictive import policies; a financial structure that puts less premium on the quarterly bottom line than does the United States; and a greater propensity for savings and capital investment in new equipment and technology.

An especially important feature of the Japanese story is highlighted here – the human resource. This emphasis is put into clear focus by a planning document of MITI. It states, "It is extremely important for Japan to make the most of her brain resources, which may well be called the nation's only resource, and thereby to develop creative technologies of its own . . . Possession of her own technology will help Japan to maintain and develop her industries' international superiority."[3]

The nation, its industries, and institutions are highly adept at mobilizing people and at putting both physical and intellectual labor to optimal use. As a result, one finds a level of technical competence spread evenly throughout the society. In industry, for example, this competence stretches from the highest managerial levels, where most managers are trained in engineering fields, down to the worker level, where the average worker has matriculated from a secondary school system that prides itself on producing the most literate population in the world. Extensive reporting of the famed "quality circle" phenomenon often ignores a key point that harks back to the educational system in Japan. The success of quality circles is due in large part to the high quality of the participants, an attribute that would be difficult to match in equivalent environments in the United States.

This emphasis on people permeates all levels of social activity. It starts with the belief and practice of "nemawashi." Yoshimatsu Aonuma, professor of industrial sociology at Keio University in Tokyo, explains this Japanese term: the meticulous maneuvering necessary to lay the groundwork for group decisionmaking and social consensus. Originally denoting the careful digging around the roots of a tree in preparation for transplanting, nemawashi has come to refer to the process of consensus building so typical of Japanese organizations.[4]

Consensus building is the glue that makes Japanese decisions stick. At the same time, the process of building consensus provides an energy that channels the flow of information, from the bottom upward, toward a specific mission, defined from the top down. "In Japan," says Aonuma, "consensus is generally formed by someone taking the lead, and others aligning with this position or idea." This process takes place largely behind the scenes, and is thus not readily visible to most outsiders. What many American executives see is the final parts of a laborious process. They see the tree after it is planted or, in business terms, they see only the outcome of a decision, for example, to market 64K chips or to launch a government-sponsored fifth generation computer project. Individual responsibility for various parts of the decisionmaking is lost to the inexperienced foreign observer. "You can't tell where Fujitsu begins and the Japanese government leaves off," says Edward Lesnick, assistant to the president at Wang Laboratories.[5] That is nemawashi working at its best. Therein lies a primary foundation from which Japanese strategy derives its dogged determination.

If consensus building among the participants provides Japanese strategy with power, what provides its shape? Many different explanations have been put forward: Theory Z management style (William Ouchi), the Seven-S's (Athos and Pascale), quality circles, and so on. For our purposes of trying to understand the challenges to American high technology industry, three principles concerning how one maximizes the human resource are key to the effectiveness of Japanese strategy in the electronics and communications technologies. These are *education and a life-long perspective, information gathering,* and *concentration* of effort.

Education and a life-long perspective. Japan is not only one of the most literate societies in the world, but also one of the most technologically literate. In the same way that countries like the United States have tended toward law and business degrees, Japan has been especially strong in technological subjects. On a per capita basis, Japan produces nearly triple the number of electrical engineers than does the United States. In 1980, Japan graduated nearly 87,000 engineers, compared with about 78,000 engineering degrees in the United States, where the national population is twice as large. Nearly all of the Japanese engineers will go to work in civilian sectors whereas approximately half of the American engineers will work on defense-related projects.[6]

JAPAN VERSUS THE U.S.

The Professional Numbers Game

Number of professionals per 10,000 population	U.S.	Japan
Accountants	40	3
Lawyers	20	1
Engineers	25	35

Electrical Engineering Graduates (B.S., M.S., Ph.D.)

	Number of graduates		Percentage Change	
	Japan	U.S.	Japan	U.S.
1971	15314	17359	+10.0%	—
1972	16052	17131	+ 4.8%	-1.3%
1973	17165	16999	+ 6.9	-0.8
1974	17419	15520	+ 1.5	-8.7
1975	18040	14331	+ 3.6	-7.7
1976	18258	14214	+ 1.2	-0.8
1977	19257	14290	+ 5.5	-0.5

One reason for the larger numbers in Japan is the social importance placed on the engineering degree. The National Science Foundation reports that in Japan an engineering degree is treated as a ticket to success "in much the same way the M.B.A. is viewed in the U.S." This in effect pulls large numbers of the most talented students toward engineering early in their education. In 1980, reports the National Science Foundation, one out of five bachelor's degrees and two out five master's degrees in Japan were granted in engineering fields, compared with one out of twenty for each level in the United States. One half of Japan's elite senior civil servants hold engineering-related degrees. In industry about one half of all directors have similar qualifications.[7]

The Japanese are working hard to upgrade an educational system that still is criticized for overstressing rote learning and underemphasizing creativity. Traditionally the universities decide which graduates will work in which companies; and once employed, the highly trained graduate is led through a career that resists specialization. Thus, in many ways one finds a missing entrepreneurial imagination that is normally associated with a more highly specialized interests.

Because large Japanese companies are committed to life-long employment policies, many have found it necessary to set up company-sponsored institutes. For example, over 5,000 engineers have studied at the Hitachi Institute of Technology since it was founded a decade ago. "We have lots of time to train engineers in our lifetime employment system," says Tetsutaro Iritani of Sony. "That means today's young engineers will become excellent engineers in the future."[8]

The system of lifelong employment that prevails in large companies (it is nonexistent in most small firms) and the loyalty to a company are a two-sided feature of the Japanese economy. For corporate planners it allows far greater flexibility in reallocating human resources to changing strategic priorities. This is often accompanied by a corresponding responsibility to train and retool employees for emerging new technologies and market opportunities. Such ongoing educational activities coupled with the longevity of the employment contract are instrumental in capitalizing on another Japanese quality – attention to detail and to perfection. This makes it possible for companies to hone manufacturing processes and products to a fine edge, something far harder to achieve in the normal American industrial environment. In the United States a constant and often rapid turnover of technical personnel works against perfection and continuity in product development programs. This problem is especially bothersome to high technology industry where, in some cases, it can take five years or more to establish efficient volume production.

On the other hand, the greater mobility of U.S. corporate employees as well as the superior flexibility of equity investment capital nurture the formation of new entrepreneurial ventures. These, plus the freedom to experiment in less structured settings, are largely responsible for the high rates of American technological innovation. In contrast, the Japanese show little natural proclivity to join new enterprises of the maverick entrepreneurs. Instead, they excel in process-oriented manufacturing technologies that in turn lead to higher quality products at highly competitive prices. If there is concern about the future in America, it is because when the curve of technical innovation begins to flatten, the Japanese may be best able to capitalize on their manufacturing competence and efficiency.

Information Gathering. Another facet of Japanese strategy is a seemingly insatiable appetite for information — a passion that is dependent on a highly motivated quest for knowledge that may affect one's company goals. No decision is made without exhaustive attempts to gather all possible sources of pertinent background material. What might be viewed in American corporate culture as a fetish is viewed by Japanese executives and government planners as a precondition to decisionmaking. Such intelligence gathering is a task shared among government agencies, corporations, and the ubiquitous Japanese trading companies scattered throughout the world.

JAPANESE INTELLIGENCE GATHERING: An Insatiable Appetite

The Samuel Johnson Case

In 1974-75, an American business research firm seeking new clients invested $120,000 in advertising. The catchy copy quoted from the eighteenth century English philosopher Samuel Johnson:

> "Knowledge is of two kinds. We know a subject ourselves, or we know where we can find information upon it."
>
> — Samuel Johnson, 1775

The ads, appearing in the *New York Times*, the *Wall Street Journal, TIME* magazine, and other business publications were aimed at attracting prospective American corporate clients for research services.

"We drew 350 to 400 responses," says Christopher Samuels, a Principal of the firm that placed the ads. "Of the American respondents, not one became a steady client. But much to our surprise — as we had not intended this result — we ended up with seven or eight Japanese companies out of the twenty to twenty-five who answered our ad!

"We learned two things from that experiment in advertising," says Samuels, now managing director of the Center for Strategy Research in Cambridge, Massachusetts. "The first is that Japanese companies deliberately look for an excuse to gather information. Japanese management appears to have little difficulty in justifying ongoing research into markets, products, competitors, and technologies. This has not been uniformly true of our experience with American corporations.

"Second, unlike American firms that will invest in massive research only when the crisis is upon them, the Japanese will be there six to nine months before the crisis. They invest in slow, low profile research. The difference is analogous to long term preventive medicine for a cardiac patient instead of emergency heart surgery. The costs and danger to the patient for the first are significantly less than for the second.

"When the crisis ultimately hits," continues Samuels, "many traditional channels and sources of information close down. By that time, however, the Japanese have the information they need."

Source: Interview with Christopher Samuels, January 1982.

Just as an engineering degree can confer social status, so can research denote honor. For many American middle-managers, a personal assignment to conduct research on a competitor would be viewed as a misuse of expensive managerial time, possibly even auguring demotion. This clearly is not the case in Japanese companies, where great importance is attached to such tasks. The Mitsubishi conglomerate, for example, has a self-contained Research Institute in Tokyo with 400 employees. Nissan Company's New Jersey office retains twenty-five consultants to feed it information on a continuous basis. The twenty-five consultants are contributors to a composite picture that emerges not only for Nissan management in New Jersey but also for the rest of the company, worldwide if necessary.

Concentration. Two examples show how the Japanese have learned the importance of concentrating limited resources on selected projects. The first illustrates a policy of selecting and supporting "sunrise" industries, such as robotics, in which Japan has taken a commanding lead, or fiber optics. The second shows the leverage of limited resources on focused and sustained research, such as the Fifth Generation Computer Project funded jointly by the government and industry. Behind this reasoning is a realization that valued resources such as scientists and engineers must be maximized and that to spread them too thinly into too many directions would be self-defeating.

For many years, Japan's Economic Planning Agency has successfully pursued a policy of channeling support to rising sectors of the economy and letting languish those sectors that are weak or declining. One way this is carried out is through the banks. "When an industry appears on the endangered list, this is a signal to the commercial banks. At the next credit squeeze, don't expect any overdrafts," writes the *Economist.*[9] Some older industries already on the endangered list are general textiles, rubber goods, plywood, stringed musical instruments, and general shipbuilding.

In contrast, when the U.S. government intervenes (which is seldom), it usually bails out failing companies, amidst great controversy. When the Japanese government intervenes (which is often), it usually supports growing, successful companies. The effect is to concentrate people, capital, and motivation in those areas most likely to ensure economic growth.

Computers, communications, and electronics are leading industries presently on the Japanese support list. One of the most successful in this regard has been robotics – the manufacture and use of computer-driven assembly line workers who never take coffee breaks, lunch breaks, sick leave, or vacations. There are a reputed 45,000 robots in operation in Japan compared to one ninth that number in the United States. Even by stricter American definitions of what constitutes a robot, the 14,000 currently at work in Japanese plants outnumber the 4,100 in U.S. factories.

According to industry analysts, almost 32,000 Japanese robots, worth in excess of $2 billion, will be produced in 1985. The best estimates show American production capacity at one fourth that rate. Such a competitive headstart implies not only control of a product market but a powerful impetus to the productivity and manufacturing capacity of entire industries.[10]

THE INTERNATIONAL ROBOT POPULATION

Programmable Robots in Operation

Country	Robots
Japan	14,000
United States	4,100
West Germany	2,300
France	1,000
Sweden	600
United Kingdom	500

Source: "New in Japan: the Manless Factory," *New York Times, December 13, 1981.*

The potential for growth in robotics is attracting major players. Even IBM is getting in on the new market. It announced production of its first robot, the IBM 7535, early in 1982. Originally of Japanese design, the

7535 is being manufactured by a one-year-old Japanese firm, Sankyo Seiki Manufacturing Company, Ltd. IBM has also announced its intent to produce an even more advanced robot, the RS1. Based on a decade of work at the company's Yorktown Heights Research Division and programmable with a personal computer, RS1's programming language (AML) is probably as important and novel as the hardware product itself.[11]

By encouraging industrial priorities, Japan is able to concentrate managerial and worker effort in those sectors that are most likely to succeed. As a result, outmoded managerial styles and entrenched work habits do not get institutionalized in old industries – a form of social innovation that is proving vital to Japan.

Another example of how Japan has learned to concentrate scarce resources is the highly select research and development support MITI provides to consortia of companies willing to focus on precise product goals. One of the potentially most far-reaching is the Fifth Generation Computer Project, an effort considered as important to the world of knowledge as the shuttle has been to the world of space. MITI is expected to provide $400 to $500 million to this project over the next few years.

The fifth generation computer builds on concepts of artificial intelligence and pattern recognition first developed at Massachusetts Institute of Technology. The new machine is meant to be a "knowledge processor" as opposed to a "data processor." This means it will be designed to process not only quantitative data but also nonnumerical information such as language, pictures, and images. One of its most ambitious targets is to be able to respond to ordinary human speech. By 1990, experts hope to develop a computer vocabulary in excess of 10,000 words with a minimum 90 percent accuracy rate. The immense complexity of using character-oriented Oriental languages for typewritten input gives the Japanese a special incentive to perfect speech recognition systems.

What is significant about the Fifth Generation Computer Project in terms of strategy is not so much its ambitious goals or imaginative capabilities as the fact that it integrates an entire set of decade-old substrategies of intellectual research. Quietly but systematically, MITI has been leading the development of a fifth generation machine by discouraging redundant efforts and by supporting research into its critical component parts such as pattern information processing (PIPS) or state-of-the-art very large scale integrated circuits (VLSI). In the footnotes are provided the extensive record of MITI's financial support for computer related research.[12]

Nemawashi and the three principles stress the importance placed on people and group cohesion. Education is a central feature of its effectiveness; information gathering is the backdrop essential to decisionmaking; and concentration is a corollary to efficient use of scarce resources – of which human intellect is the crucial ingredient in a knowledge-intensive economy. These are principles not commonly practiced by the United States or by many corporations.

French Strategy: Technological Nationalism First

Since the time of Louis XIV, France has nurtured an administrative meritocracy reserved for only the best minds. Consistently, educational resources were invested to ensure an inculcation of the highest standards and ambitions in the administrative elite. Oftentimes, this served the counterproductive purpose of narrowing the field of opportunity to too few bright and entrepreneurial youngsters. These characteristics, still pervasive in the culture, explain a dilemma of French policy. The best laid and intellectually sound plans are not always executed with the same degree of competence and flair as their originators idealize. One might term this problem the "Concorde Dilemma" or the art of building a magnificent plane without testing whether the marketplace wants such a product. As France faces its technological future, this dilemma will remain a significant concern of policymakers as they formulate competitive strategies.

Starting in the mid-1940s, after the close of the second world war, French administrators exercised their powers through a burgeoining central economic planning structure. The five-year central plan became a primary feature of the Fifth Republic after its founding by General Charles de Gaulle. Now in its eighth phase, coinciding with the ascent to power of François Mitterrand, the "plan" is less a schedule of quotas and specified production targets than it is a generalized statement of economic direction. Corporate entities are invited to participate with the highly persuasive sweetner of government subsidies, orders, and favors. While the linkage between government plans and corporate performance is not direct, companies are far from passive in their reactions to the plan. Indeed, they make significant efforts to successfully lobby the government agencies so that the plans reflect corporate preferences.

This collaborative process is a fine-tuned mechanism built on close working relationships between the administrative elites in industry and government. Key government administrators – especially from the ministries of Industry, of Research and Technology, and of Economics and Finance – maintain close associations with corporate executives who are peers from the same Grands Écoles. In many cases, this intimacy is so great that in American terms and under American law it would be viewed as collusion, conflict of interest, and perhaps even corruption.

Government in France not only plans its economic future in five year planning cycles but it plays the lead role in thinking out the long-term future for the economy and the society. Under de Gaulle, Pompidou, Giscard d'Estaing, and now Mitterrand, the administrative bureaucracies play a unique role as intellectual high priests seeing change in its loftier social fabric. This is illustrated by a program started by Giscard d'Estaing near the end of his seven-year administration. He ordered the convening of an international symposium on the relationship between the new information age and society.[13] A principal purpose for this colloquium was to understand social and economic trends affected by the electronics and computer age. The six volumes of findings would serve as a foundation for investment decisions by the national government. Equipped with a sense of future, the government could better coordinate its programs strategically in order to maximize economic opportunity and minimize social dislocations, such as with the declining coal and steel industries. The télématique and télétexte program to place interactive terminals in every French home and office was conceived with a dual goal of anticipating the social needs and impacts of the electronics age and of providing new employment opportunities.

The most impressive feature of the French planning process is the major role government purchases play in driving the economy. This is now the linchpin to France's bold and rapid entry into a number of high technology industries – most visibly in telecommunications, aeronautics, and nuclear energy. Under de Gaulle's administration, government demand was used as a method of reversing French dependence on American technological hegemony in military hardware. Long-term programs were launched to build an arsenal of nuclear weapons, submarines, aircraft, and missile technologies. These efforts – not dissimilar to a U.S. undertaking during the 1950s and early 1960s when companies such as IBM counted almost 60 percent of their revenues from federal contracts – yielded a valuable by-product of skilled knowledge workers and an infrastructure of high technology supplier firms.

But reliance on government purchasing power had one key drawback. France's technological progress could move only as fast as the depth of its public purse, which was under the costly stress of sustaining the last vestiges of colonial power in Algeria. Despite the limited resources at hand, the results were technologically eminent. In 1967, a watershed year for French technology and national pride, the weekly magazine *Paris Match* announced in bold letters: "FINALLY! – A New Plane – 100% French – The Interceptor SUPER MIRAGE F-1."[14] Subsequent generations of the Mirage, Concorde and Airbus technologies, Aerospatiale helicopters, successful launchings of the Ariane space rocket, state-of-the-art nuclear breeder reactors, highly sophisticated telecommunications technology, and a national software capacity second only to the United States all speak to the surprising success of French strategy.

When the first Plan Calcul was announced in 1967, public funds were invested in forcing the merger of three smaller firms into the Compagnie Internationale d'Informatique (CII). The combined work force of the new entity totalled 2,600, of whom 550 were engineers and another 700 technicians. CII would later combine forces with Honeywell Bull to form an even larger entity eventually to be nationalized in 1982. The amount of government support provided to French computer manufacturers under the Plan Calcul is given in the footnotes.[15]

The main lines of Gaullist policy, which has been consistently followed since then, were well summarized by *The Economist* in 1965:

> The French feel that importing patents and licenses cannot be a certain method of ensuring a satisfactory pattern of home production and exports in the long-run . . . At the moment, the emphasis of the campaign is on joint research projects between firms in the same sector. In the longer-run, the government hopes to promote an increasing number of mergers where French industry is too small scale. Beyond that they see the prospect of more intra-common market collaboration.[16]

This has produced companies with such unwieldy titles as Saint-Gobain-Pont-A-Mousson. Originally a glass and piping conglomerate, it diversified into electronics. It, too, was nationalized. It owns a majority interest in Compagnie Machine Bull which, in turn, controls a majority of

CII-Honeywell Bull (CHB). The parent group and CHB also bought a 30 percent interest in Olivetti.

As a stimulant to private industry, large and lucrative government orders are placed through agencies such as the Ministry of Posts, Telegraph and Telephone (PTT) with local suppliers. Forty percent of the semiconductor chips sold in France are purchased by public agencies in a local market that is expected to grow to $200 million by 1985. This public leverage is used to impose knowledge transfers through joint venture with U.S. firms. In this way, Intel Corporation of California was forced to share proprietary product knowledge in its micro-chip association with Matra S.A. and Harris Corporation. Earlier, during the Giscard d'Estaing government, similar pressures were the basis for a marriage between Thomson-CSF and Motorola and between Saint-Gobain-Pont-A-Mousson S.A. and National Semiconductor Corporation of California. The present government of France, like Giscard d'Estaing's former administration, is pushing and goading its high technology corporations to compete in world markets.

SIX FRENCH COMPANIES: Going after world markets

SAINT-GOBAIN-PONT-A-MOUSSON, 1980 sales 43.5 billion French francs (FF). Large pipe and glass group diversified into electronics. Owns 51 percent of Cie des Machines Bull, which controls a majority in CII-Honeywell Bull. Bought into Olivetti. Partner with National Semiconductor (US) in chip making.
[SGPM Nationalized 1982]

CII-HONEYWELL BULL, 1980 sales 6.3 billion FF. France's biggest computer manufacturer, owned 80.1 percent by French interests, 19.9 percent by Honeywell of the United States. Heavily supported until this year by state aid.
[SGPM owns 53 percent of CHB; agreement to reduce Honeywell (U.S.) participation from a 47 percent interest to 19.9 was reached in April 1982.]

THOMSON-BRANDT, 1980 sales 36.6 billion FF. Big electrical and electronics group spanning most industrial and professional consumer products. Its television manufacturing arm, the biggest in France, has been rapidly expanding through acquisition of tube and set-making concerns in France and Germany.
[Nationalized 1982]

THOMSON CSF, 1980 sales 22.3 billion FF. Part of Thomson group, involved in radar, avionics, minicomputers, telecommunications components, and medical electronics. Defense work is 37 percent of sales and biggest profit earner. Collaborating with Xerox (U.S.) in development of advanced computer memory devices.
[40.04 percent owned by Thompson-Brandt]

CIT-ALCATEL, 1980 sales 8.2 billion FF. Part of the big Compagnie Generale d'Electricite - CGE (1980 sales 45.8 billion FF). Outstanding success in telecommunications equipment, claims to have installed 60 percent of the world's digital lines.

ALCATEL ELECTRONIQUE (1980 sales 4.7 billion FF) is rapidly entering the office automation market.

[CGE, Nationalized 1982, owns 49.2 percent of CA, and 81.8 percent of AE]

ENGINS MATRA, 1980 sales 5.6 billion FF. Rapid growth of defense business which accounts for about half of turnover. Acquired interests in telecommunications, watches, instrumentation, and motor components. Major contractor for Telecom 1 satellite program. Joint venture with Harris (U.S.) to make microchips in France.

[National government acquired 51 percent interest in 1982]

Source: Adapted from *Financial Times*, October 1, 1980; *L'Expansion*, November 1981; *Le Monde Informatique*, February 22, 1982; and *New York Times*, April 22, 1982.

These initiatives are bearing tangible fruit for French suppliers. In May 1982, Tymshare (U.S.) announced the purchase of 560,000 low-priced terminals from Matra, a favored supplier of the government and the PTT. Earlier, Alcatel Electronique sold one quarter of a million terminals to Source Telecomputing, a division of the Reader's Digest Corporation. And Thomson was booking an order for 35,000 business terminals from GTE – in dollar terms the largest order registered to that date in the telephone-based terminal sector.

The Mitterrand nationalizations are one step further on the continuum of government efforts to move the French economy to a position of world competitiveness. Takeovers of private firms by the government in early 1982 gave it direct control over one half of the French informatics industry. Late in the fall of 1981, Mitterrand's resolve was evident in the announcement by his minister of research and technology, Jean-Pierre Chevènement, that the research and development budget would be increased 29 percent to an amount equal to $4.4 billion in 1982. By 1985, the aim is to devote 2.5 percent of the GNP to research and development, up from 2 percent in 1981. An immediate effect of the increase will be to expand faculty and research positions in universities by 1,800.

Recruiters were sent outside of France to seek candidates for new faculty positions. At MIT, aggressive employment offers were being made soon after the policy announcement.[17] A striking result was the recruitment of Nicholas Negroponte, a noted professor of computer graphics, to be director of the newly created World Center for Microcomputer Science and Human Resources. His chief scientist will be an MIT colleague, Seymour Papert, famed for his pioneering work in developing a computer-based education system known as LOGO. Both men are on leave from

MIT. In addition, the French center signed on as consultants Atari's chief scientist, Alan Kay, and Carnegie-Mellon University's director of its robotics institute, Raj Reddy.[18] The Center also counts among its board members the artificial intelligence expert Marvin Minsky and former MIT president Jerome Wiesner.

As France faces the coming decades, telecommunications have the greatest promise of capturing large shares of the world market. Government officials ambitiously project an expansion of their domestic industry from a 19 percent share of world sales to more than 50 percent of total sales by 1987. One sector of the world market aggressively courted by France is Third World telecommunications purchases. The World Center is positioned to play an important role in cultivating these markets in part by providing technical education to prospective buyers. Plans are already envisioned for such programs in Senegal, Kuwait, Ghana, the Philippines, and Saudi Arabia. It is not surprising that the center's chairman is Jean-Jacques Servan-Schreiber, an ex-cabinet officer and author. His recent book, *The World Challenge*, stressed the need for just such a center and pointed to the OPEC nations as key partners in future French projects.

FRANCE's WORLD COMPUTER CENTER:
A Door to Third World Markets?

Founded late in 1981 at the suggestion of Jean-Jacques Servan-Schreiber, the center is a favored project of President François Mitterrand. With a start-up budget of 100 million French Francs (or about $16 to $17 million at 1982 exchange rates), a central goal of the center is to devise education and training programs on personal computers for use in developed and developing countries, with emphasis on the latter.

The importance vested in the center is evident not only in the speed with which it was established, (several months) but the high level ranks of five of its board members — all cabinet individuals representing Interior, Post, Telegraph and Telephone, Industry, Culture, and Research and Technology ministries.

Its chairman, Servan-Schreiber, has the highly unusual right of reporting directly to Mitterrand, with the equivalent of ex-officio cabinet ranking status — some say because Mitterrand is courting the rightist constituency of Servan-Schreiber in the event of a fallout with the Communists. Soon after founding the center, Servan-Shreiber was cultivating new funding sources from cash-rich OPEC nations; and he quickly announced a uniquely distinguished scientific team under the direction of a U.S. computer graphics pioneer, MIT professor Nicholas Negroponte. The Center's team includes Seymour Papert from MIT, Edwards Ayensu from Ghana, Kirgsten Nygaard from Norway, Fernando Flores, an ex-minister in the Allende government of Chile, Zhisong Tang from China, Keichi Oshima of Japan, and Nabil al Yassini of Kuwait. Projects in several Third World nations have also been tentatively announced for Senegal, Kuwait, Ghana, the

Philippines, and Saudi Arabia. In France, one of the center's early projects is aimed at retraining workers displaced by robotics.

Sources: *Le Monde Informatique*, November 23 and 30, 1981;
Temps Réel, November 30, 1981;
Express December 4-10, 1981;
Washington Post, January 14 and 27, 1982.

France is also the largest supporter of the Intergovernmental Bureau for Informatics (IBI), an international organization examining national information policies and especially the development of informatics in Third World countries. Its Declaration of Mexico, signed by President Lopez Portillo, stressed the need for cooperation between industrial and developing countries to use computers for economic development and for peaceful purposes. The IBI's world computer conference (SPIN II), scheduled to be held in Havana, Cuba, will include some forty French company representatives but, due to political considerations, few if any North American company or government officials will attend.

While many Americans are reluctant to suggest that French technological expertise will undermine the U.S. economy in similar ways as the Japanese, it is evident that France has risen surprisingly fast to the ranks of a peer in a number of key high technology fields. A public program to broaden the reach of education and to increase rapidly the ranks of skilled workers in knowledge-intensive technologies, and a parallel recognition that entrepreneurial incentives are needed to encourage technological progress, is considered by planners as an essential part of the French thrust.[19] A solution to this side of the technological equation will also solve the nagging reminder of the "Concorde Dilemma".

Current French policymakers are zeroing in on well defined high technology targets. According to Ambassador Michael B. Smith, the U.S. representative to the General Agreement on Tariffs and Trade (GATT), "the French sifted through 600 industrial sectors, using 33 criteria to pinpoint 23 priority sectors. Out of these, six have been selected as target industries for special development."[20]

Ultimately, though, the real measure of success between competing knowledge-intensive economies will be found in the quality of their human resources. For Americans, education, motivation, and innovation are familiar ingredients to success. What is apparent in the United States is that the nation's ability to sustain its output and motivation of highly trained people — engineers, managers, and workers — is more than ever in question. Misguided or misunderstood priorities have shifted the nation's attention away from the single most important task: maximizing human potential through education. This is the one asset America has traditionally cultivated to its fullest and upon which much of its economic vitality has rested in the past, and should rest in the future.

Chapter 4

A NEW SCARCITY: HUMAN RESOURCE SHORTAGES

IF ANALYSES of present trends are correct, the prognosis is not good for the United States. The discrepancy between the demand for people in high technology industries and those supplied by educational institutions is growing severe. According to the president of Princeton University, liberal arts candidates are being turned out at a rate four times greater than demand.[1] At the same time, according to a number of professional engineering, educational, and industrial associations, electrical engineers and computer scientists are in demand at levels that far exceed the supply coming from colleges and universities.[2]

This mismatch has been building since the early 1970s, a decade that saw a doubling of the number of graduates from law schools while the output of electrical engineering degrees remained flat. What is surprising here is not the explosive growth in lawyers (who may well be needed, although the Bureau of Labor Statistics projects a 20 percent oversupply by 1985) but the stagnating undersupply of engineers, despite a continuing high level of demand from industry.

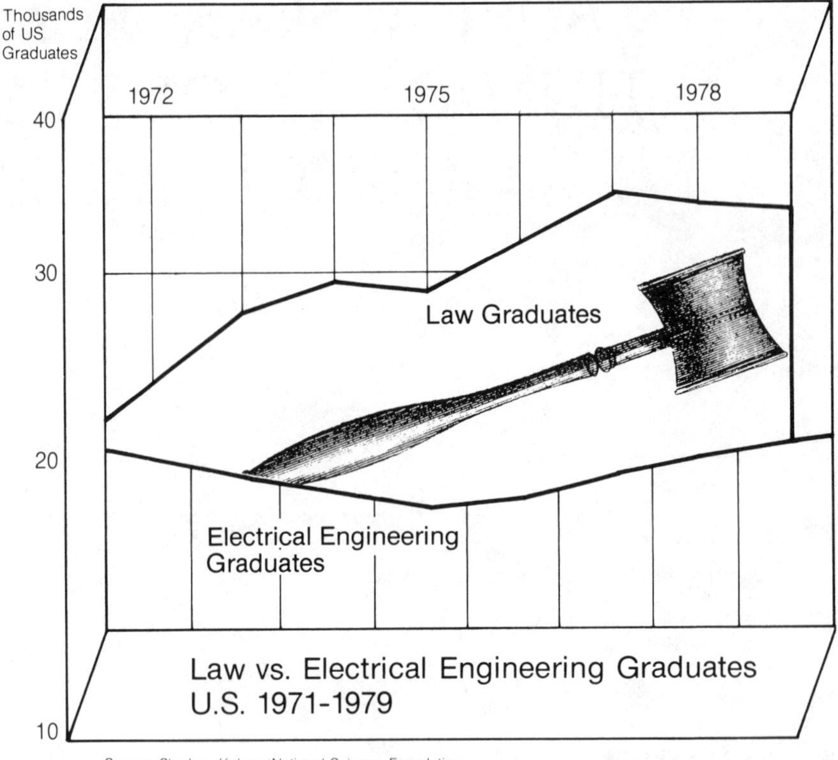

Source: Stephen Kahne (National Science Foundation

If it is true that the present and projected supply of a critical resource remains constricted in the face of market demand, we need to find out why. In this chapter, we examine the causes, extent, and likely duration of shortages in two fields of particular importance to the high tech electronics industry – electrical engineering and computer science.

Supply and Demand: The Facts

No significant long-term relief is foreseen in the national demand for electrical engineers and computer scientists. Unless new actions are taken, no dramatic increase in supply is to be expected over the next five years or more. This is the basic conclusion reached by studies sponsored by a number of professional societies and industry associations including the American Association of Engineering Societies, the National Academy of Engineering, and the American Electronics Association (AEA).

Setting aside computer scientists where there is less history, let's look at the case of electrical engineers in order to understand the dynamics and dimensions of the shortage. According to the Bureau of Labor Statistics, the employment growth of the electronics industry was 5.1 percent from 1959 to 1969 and 1.7 percent to 1979. The projected growth for 1979 to 1990 is 2.6 percent per year. This figure includes both the slower growing radio communications sector as well as the faster growing computer and electronics sector. Considering the build up of military procurement needs, it is not hard to imagine that the demand in the 1980s will be more like the 1960s when it averaged more than 5 percent.

The net growth in the supply of electrical engineers turns out to be significantly less than these historical or projected demand figures. New graduates (in 1980, 17,777 degrees, less 20 percent who enter nonengineering professions) plus immigration (in 1980, just under 1,000) will add 7.7 percent new members to the electrical engineering workforce. The electrical engineering workforce is estimated at about 260,000 and approximately 206,000 work in industry. Since the pool is depleted each year due to death and retirement (one percent) and those who switch to management positions (estimated at 5 percent), the net growth is only about 1.7 percent per year. This is far less than the minimal 2.6 percent rate of growth in demand projected by the Bureau of Labor Statistics.[3]

The American Electronics Association, after surveying its member companies, undertook an analysis that arrived at similar conclusions.[4] Indeed, the AEA report concludes its report by saying: "To meet just the needs of electronics, education must triple its output of EE/CS engineers each year for five years."[5]

In selected areas, the shortage is especially severe. For example, in 1979 the need for new computer-related Ph.D. graduates for positions in industry and academia totaled 1,300.[6] Yet only 190 doctorates in computer sciences were granted that year for the entire nation. And in 1980, that figure dropped to 159. (In the notes at the back of the book are given the data on the number of engineering graduates for the years 1970-1980.)[7]

A regional estimate of manpower needs in California, completed by the AEA in 1980, indicated a demand by the 3,500 electronics firms in the state for 4,111 new electrical engineers and computer scientists (EE/CS) per year through 1985. These numbers did not account for a possibly immense employment surge from rising defense expenditures, anticipated to add an overall total of three-quarters of a million jobs to the state over a four-year period. In the AEA's words, "juxtapose these demand figures against the 2,909 EE/CS graduate degrees (all levels) awarded by all public and private colleges in the state in 1980, add the issue of mobility [high California housing costs are a deterrent to potential employees from out of state], and the problem's magnitude grows."[8]

A similar analysis of regional supply and demand completed by the Massachusetts High Technology Council reached the same conclusions, showing in this case a shortfall of 1,000 electrical engineers a year for the "Route 128" firms clustered around Boston. In the Phoenix region of Arizona, an industry/academic consortium at the University of Arizona that includes Honeywell, Motorola, and some forty other companies estimates that they need 1,500 new engineers annually.[9]

"VLSI" SHORTAGES: Two Thousand Openings for Two Skills

There are approximately 264,000 electrical engineers in the labor force in the United States. Of those, some 6,000 are employed in the semiconductor industry. Two critical skills pace the growth of the semiconductor industry's most important project — "VLSI," the development of Very Large Scale Integration techniques. Those two skills are semiconductor process engineering and semiconductor design.

A study made in 1980 showed there were approximately 3,750 people with such skills in the entire free world. Of that number, approximately 2,450 are located in the United States. But the demand for these two critical semiconductor skills is estimated at nearly 4,500. The shortage is thus expected to be approximately 2,000 in the United States by the end of 1985. These estimates were made by a spokesman from Motorola, presently the largest merchant supplier of 64K VLSI circuits.

Source: I. R. Saddler, Motorola, Inc., Symposium Proceedings 1981, University-Government-Industry Microelectronics Symposium sponsored by IEEE, NSF, & the International Society for Hybrid Microelectronics at Mississippi State University.

Stephen Kahne of the National Science Foundation sees a pronounced shortage of serious, even crisis, proportions. He has reported on this issue in a July 1981 article entitled "The Crisis in Electrical Engineering Manpower":

> It is no longer an open question whether the shortage of electrical engineers in the United States is or is not a crisis: it is. Sooner or later every U.S. industry dependent on electrical engineering will be affected — and there are more such industries now than ever. Indeed . . . new industries, previously unaffected by electrical engineering in any significant way, are the hidden factor that invalidates traditional market need for electronic specialists in sectors of the economy that never before employed them.[10]

AN EXPLOSION IN NUMBERS:

The International Computer Population Reaches 10,000,000

In 1975 there were about 500,000 computers, big and small, installed throughout the world, two-thirds in the United States. Five years later the number jumped to 3,100,000 worldwide, and by 1985 there will be in excess of 10,000,000. About half, or less, will be in the United States.

Ten years from now, most offices and many homes will have their own interactive data terminals; the quantities will be measured in the tens of millions. France alone had a goal to install 30,000,000 terminal/telephone units in every home and office by 1990. The project has been scaled down by the Mitterrand government, but not abandoned.

The 64K semiconductor chip market, which is currently a $100 million business worldwide, will expand to twenty-five times that size in five years!

The anticipated growth forecasted by the Bureau of Labor Statistics in discrete sectors such as computer specialties is very large. For the period 1978 to 1990, almost 2 million new jobs will be created. In 1980 and 1981, the Bureau shows demand for B.S.-level graduates in computer sciences was 20 to 40 percent above the supply.

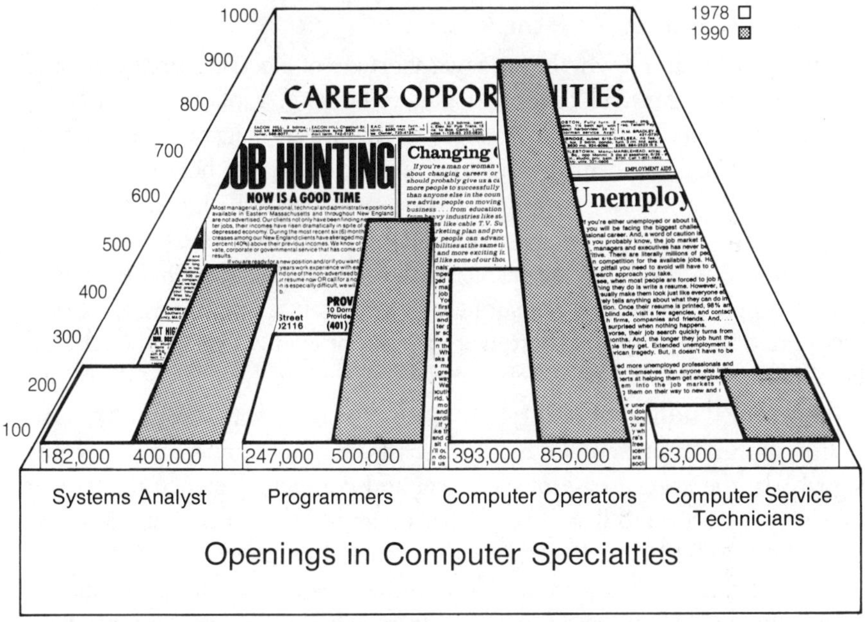

Source: Bureau of Labor Statistics

Cyclical or Long Term?

The projections of supply and demand for electrical engineers and computer scientists, or for any engineers, needs be taken with some skepticism. Critics are quick to point out the weaknesses of employment predictions based on data that are hard to measure and intentions that are hard to quantify. Measuring and forecasting employment needs has been and remains a contentious subject. Wrong so many times, any extrapolations of numbers are viewed as highly suspect.

Professor J. Herbert Hollomon and Dr. Marvin Sirbu of MIT contend that the historical record shows demand for engineers to be cyclical. Their models suggest that the engineering supply-demand problem typically moves in short term cycles of five years or so and is thus self-correcting. The statistics confirm this recent pattern if all engineering professions are aggregated, and viewed against backdrop of engineering layoffs when NASA scaled down its space operations, it rings even more true. Yet even Sirbu is candid in admitting that "these ideas are based on generalized information about *all* scientists and engineers. We did not attempt to look at subsectors such as electrical engineering or computer sciences," as distinct from aerospace, for example.[11]

Only time will tell whether today's shortage of electrical engineers and computer scientists will become a glut tomorrow. Hollomon and Sirbu have been predicting that an oversupply of engineers is just on the horizon, but for electrical engineering and computer sciences, the horizon seems to stay in the distance. Because there has been a structural shift in the economy toward knowledge-intensive industry that depends on EE/CS graduates, the dip is not likely to occur in the foreseeable future. Of course, almost no profession has an insurance policy against unemployment, so it would not be surprising if there were ripples or even waves in the EE/CS employment picture. But as best we can see, high tech industry is not just a ripple but the major part of a rising tide.

For all their faults, employment forecasts can function as early warning signals. In the special case of electrical and computer science manpower needs, they raise a red flag. The picture today is different than it was in the late 1950s and 1960s when the defense-fueled engineering employment cycle fizzled to a halt. Current projections are based on a steady economic trend toward broadly based increases in new technologies, as many companies continue sales growth rates of 25 percent a year and more. The pressure is building to keep the pipeline of trained engineers flowing smoothly. As international competition intensifies, and as defense spending increases, the situation will become even tighter.

For education, this pressure is uncovering four possible problem areas that could have serious qualitative and quantitative impacts on the future of the high technology industry — faculty shortages, capacity problems, declining high school standards, and an underutilization of available technical talent.

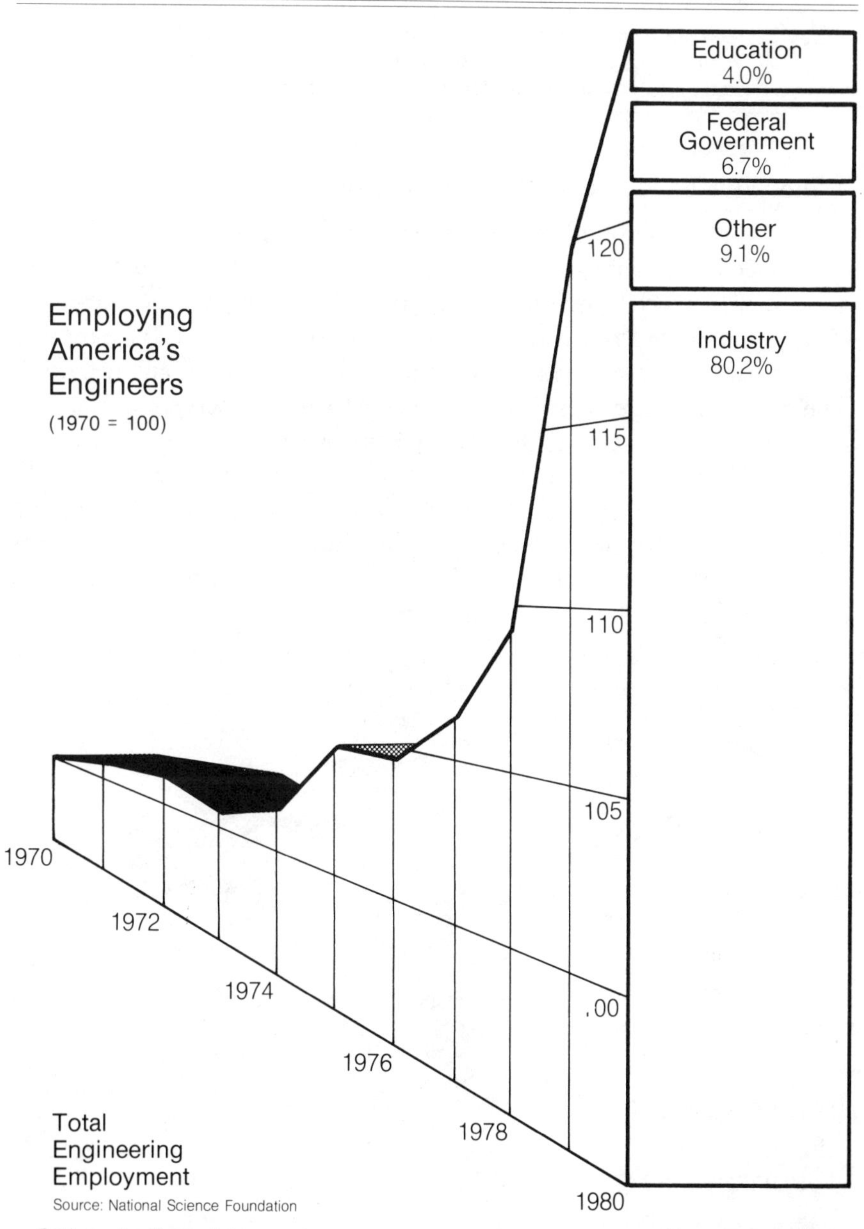

Of the scientists, eight out of ten are computer specialists, chemists, mathematicians, or geoscientists and the other two out of ten are in the social sciences. Of the engineers,

25 percent are electrical engineers; 21 percent mechanical; 11 percent industrial; 10 percent civil; and the remaining third are chemical, aeronautical, and other specialties. A relatively high number of Ph.D.s, 14 percent of the overall total, are employed by the federal government.

Source: Betty M. Vetter, Executive Director Scientific Manpower Commission, testimony to Subcommittee on Economic Stabilization, The House Banking Committee, July 24, 1981; and *Manpower Comments*, July/August, 1981, page 5.

Shortages of Present and Future Faculty

At the heart of the problem is the shortage of present and future faculty. Many faculty positions are going begging in engineering and some science departments of colleges and universities. According to the American Council on Education, in the 1980-81 scholastic year, 1,583 positions were vacant in the nation's 244 accredited departments of engineering.[12] The overall shortage is estimated between 10 and 15 percent of the total full-time faculty pool. In EE/CS departments, the shortage reaches 16 percent. While there are few statistically precise measures, Daniel Drucker, president of the American Society of Electrical Engineers (AASE) and dean of the College of Engineering at the University of Illinois, estimates that nearly 50 percent of the faculty positions in solid-state electronics, computer engineering, and digital systems have not been filled.

One of the reasons advanced for the lack of engineering faculty is the

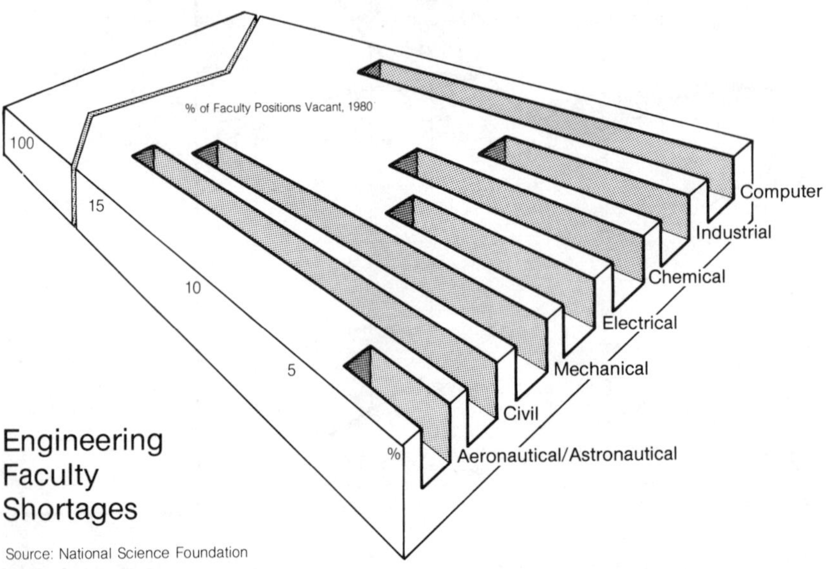

Engineering Faculty Shortages

Source: National Science Foundation

inadequacy of academic salaries. A professor of computer science can earn more in industry — anywhere from 25 to 100 percent more — to compete with the status and prestige normally associated with academic life. For bright young professors in their thirties, the discrepancy can be even greater, running as high as 200 percent or even more when the alternative to teaching is founding a new company.

Market forces are polarizing salaries to such an extent that the option to become a professor or instructor in a university is now looking more and more like a financial chastity vow. And the fewer faculty there are, the more onerous is the workload for those that remain behind to teach. Since student-to-faculty ratios are increasing, the average engineering department faculty member has a teaching burden 40 percent greater than ten years ago. One impetus for the founding of the Wang Institute of Graduate Studies in 1979, discussed in more detail in Chapter 6, was the recognition that high salaries are essential to maintaining high standards of education. At the Institute, salaries range from $60,000 to $100,000 with an aim of being dollar-for-dollar competitive with industry. Such rates are double to triple average rates in most traditional universities.

The problem is not only a present shortage exacerbated by the salary issue, but a future one as well. The number of graduate students at the

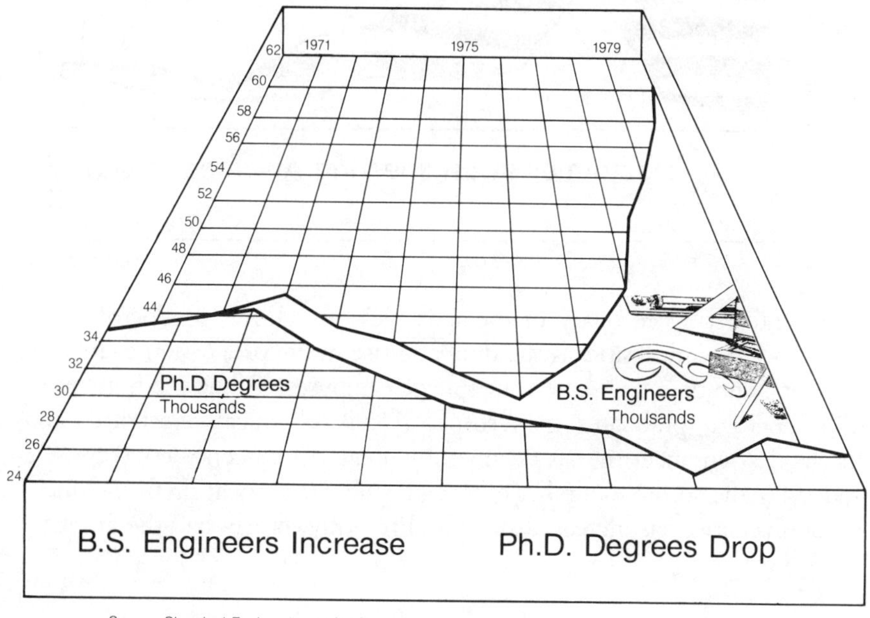

Source: Chemical Engineering 11/16/81

master's and especially at the doctorate level is declining. In 1970, there were 4,150 master's degrees granted in electrical engineering; by 1980 the number declined to 3,740. This 10 percent reduction seems small compared to a 48 percent slide, from 873 to 523, in new electrical engineering Ph.D.s over ten years. (See notes at the end of the chapter for data on the numbers of EE/CS graduates for the years 1970 through 1980.)

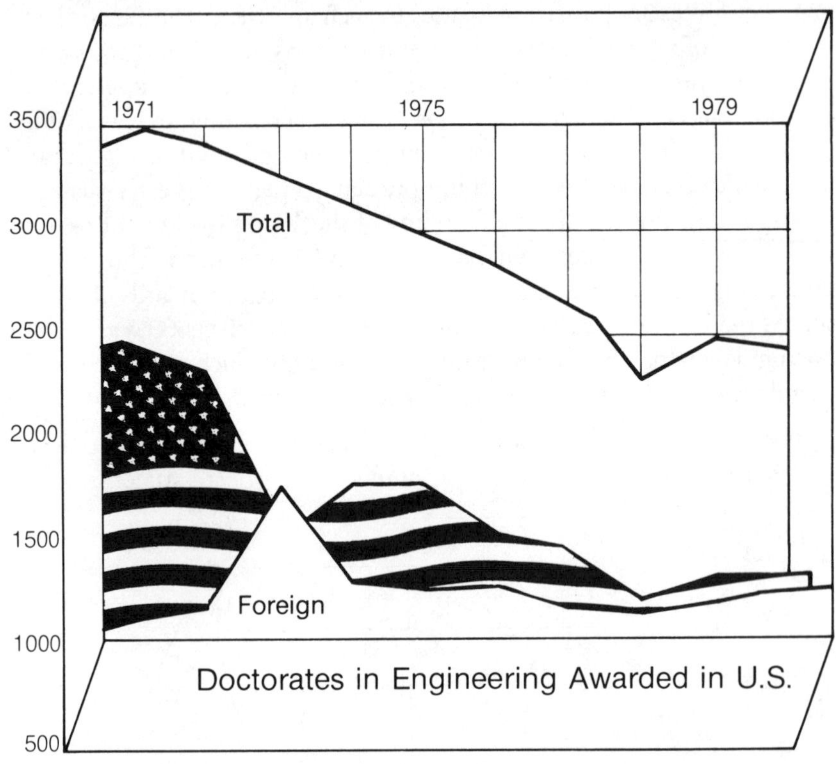

Source: National Research Council

An interesting corollary of the shortages of engineering doctoral students in U.S. universities is the dramatic rise in the proportion of foreign students who populate American college campuses. While the number of U.S. doctoral graduates in aerospace, civil, chemical, electrical, and mechanical engineering has declined by about 40 percent between 1972 and 1980, the absolute number of foreign students graduating from American universities has increased steadily. Presently, nearly half the graduating numbers are foreign citizens compared to about 25 percent in 1972. Some stay in the United States as full-time residents while others do not.

It is ironic that many who wish to stay find it difficult to be granted working permits by the Immigration Service. As of late spring, 1982, legislation had been introduced that would require foreign nationals to return home for two years after receiving their degree. This would further exacerbate an already tight situation.

The significance of the foreign student dilemma is not that we are training our future competitors (which may or may not be the case) or that foreign students are getting a free ride (which they are not), but quite simply that, like medicine before it, engineering cannot rely on American students alone to meet its domestic needs.

One of the reasons for the decline in the number of American EE/CS graduate students is, of course, also economic. When industry is offering generous starting salaries to recent graduates, it is increasingly difficult to justify staying longer in school, especially when tuitions are high. This is of course a partial explanation of why there are so many foreign graduate students who occupy the places left vacant by American candidates gone to industry.

Since the majority of future faculty members are drawn from the ranks of doctoral graduates, the ultimate effect of these declines is to reduce the pool of new faculty candidates who will train the next generation of researchers. One shortage feeds upon another. Or as the AEA put it, applying a metaphor of an agricultural era to an information age, "industry is eating its own seed corn."[13]

THE CARNEGIE MELLON CASE: One Shortage Produces Another

In 1979, Professor Floyd B. Humphrey accepted an invitation by Carnegie-Mellon University to head up its department of electrical engineering. Once arrived, he faced a faculty shortage of one third the overall need. According to the *New York Times:*

> Humphrey found that of a masters-degree class of 60 students last year, only seven stayed on to work on a doctor's degree. He feels his department needs a minimum of 20 to thrive. Without enough graduate students, there are not enough people in the department to teach the undergraduate courses. The faculty members' research projects wither, because it is the graduate students and their research which helps sustain the overall effort. Without the necessary level of research activity, the money from Federal and private research grants will decline. The

exodus of faculty, already under way because of the lucrative pay available in industry, will accelerate. The research grants will decline further, making it ever more difficult to attract new faculty and students.

Source: *New York Times Magazine*, June 28, 1981.

Rising demands from industry seem to have touched off a downward spiral in engineering education. A shrinking number of faculty members means fewer projects to sustain graduate students, and fewer graduate students means fewer future faculty. As the number of people declines, a secondary spiral is touched off: research also declines. As research winds down, so do funds to support it; and as funds are withdrawn, research withers further. This is the age-old problem of robbing Peter to pay Paul. Industry gains while universities lose. But in the long run, industry will lose as well if nothing is done to reverse the trend.

Surely this is not the first time that the balance between industry demands and educational supply has been upset. Why do the normal adjustment mechanisms not seem to be working, or at least not working fast enough? The reason is the difficulties American colleges and universities face in expanding the capacity of engineering education. In past times of abundant growth, this could have been quickly resolved. But today, with budgetary cutbacks and no growth, education faces special problems. What is the educational capacity issue, and how might it be resolved?

What Limits Expansion?

To understand the capacity issue, one must ask whether U.S. universities could increase the output of EE/CS graduates, even if they wanted to. If the answer is no, the problem is capacity. The major question then becomes: what critical resources are limiting capacity and who can supply them? The usual constraints are faculty (which, as we have just seen, are scarce in this field), money, facilities and equipment, and a supply of able students. Which of these factors, or combination of factors, are relevant here?

Before proceeding, it is interesting to note that the capacity issue affects some universities like Stanford, MIT, and other top schools differently than most.[14] Stanford's engineering school has currently set a capacity cap at 1,800 engineering students. "We already broke through the earlier 'cap' of 1500," says Engineering Dean William Kays. "I don't foresee that we'll expand that number significantly again in the coming years."[15] The cap on growth emanates from the university president's office. Concerned about the balance between engineering and other departments, President Donald Kennedy does not want one sector of the university to grow unduly large, particularly if that growth were to be at the expense of other faculties. Thus the perception of educational mission plays a role in limiting capacity. To the extent that institutions see their role as preserving a balanced curriculum, further growth in one area may be constrained by the desire to move as a university rather than as a department.

At MIT, whose primary mission is technology education, a salient bottleneck is a question of maintaining quality. Already, nearly one-third of MIT students are going into electrical engineering and computer sciences. Graduate student demand remains strong. The head of the EE/CS department, Professor Joel Moses, says that there were more than 600 graduate school applicants in the past year, and that these "were the best we've ever seen."[16] However, according to him, they tended to apply to MIT and other top schools, but nowhere else.

At MIT, strong faculty opposition arises when any move is made that is perceived to lower, or threaten, the traditional commitment to producing the highest possible standard of engineering graduate. For example, the school presently requires not only doctoral candidates but also master's students to write a graduate thesis. If this requirement were dropped (as many schools, including Stanford, already have done), more faculty time would become available to allow a larger number of master's degree students. But removing a thesis as a requirement for completing a master's degree in engineering has met with considerable opposition on the grounds that it is a step toward reducing quality and lowering standards.

For the majority of colleges and universities, the first and most important problem is money. Northeastern University and Boston University, two of the largest universities in the Boston area and only a few blocks and a bridge from MIT, would expand their engineering capacity with little hesitation if sufficient money were available. But the financial plight of these two institutions, like that of most institutions nationwide, aggravates the problem of producing more engineers. Inflationary pressures have escalated campus costs at rates exceeding any budgetary increases. For private schools, endowments have been decimated by inflation and stagnating stock prices, forcing operating losses to be made up from cash gifts that once were a key source of new capital. The bottom line for engineering and other departments means less money for salaries, fellowships, expenses, and laboratory equipment.

Public institutions are subject to budgetary strains that are similar and in some cases more severe. The total state government allocations for higher education declined in real dollars by 4 percent from 1979-1980 to the 1981-1982 fiscal year. Some states registered 17 percent reductions in budgets. Washington, Oregon, Idaho, Michigan, Ohio, and Pennsylvania have recently suffered the greatest declines in state appropriations.[17]

California, long vaunted for its extensive and superior public system of higher education, has not fared much better. The University of California at Berkeley attempted to launch a microelectronics research center in 1981. The start-up operating budget for equipment and administrative costs was set at $4 million. The legislative allocation, however, was only $1 million, far from enough to fund a major research center. Victim of a 1981 state deficit of $3 billion, the university has little prospect of sufficient state government support for the full operation of its research center.

Recent federal government cutbacks have also significantly reduced appropriations for research and education of the type that could help alleviate engineering capacity problems. National Science Foundation support for science and engineering education was earmarked for a drop from $70 million to about $10 million in one year.

While the greatest impact of budgetary restrictions is to reduce inflation-adjusted salaries for new faculty and salary increases to retain present faculty, financial shortages also limit capacity in terms of equipment and facilities. The American Society of Engineering Education sampled the age of teaching equipment in several universities and found much of it to be twenty to thirty years old. Equipment to teach the new "growth technologies" like microelectronics existed in only a small percentage of the nation's engineering schools. The inventory of computer laboratory equipment is thus badly out of date. A new generation of computer scientists is being brought up on an obsolete generation of computers.

UNIVERSITY OF MASSACHUSETTS: Shortage of Equipment

At the University of Massachusetts, Professor George Zinsmeister of the Department of Mechanical Engineering teaches a course in computer aided design (CAD/CAM) to about 35 students and expects more than 100 soon. But even the current enrollment in his class has great difficulty getting time assigned to them on the graphics terminal needed for CAD/CAM. In his words "it used to be that when a company hired a university graduate, it was hiring the latest in knowledge. The university was an important source of the infusion of new knowledge. But not any longer. The university was ahead; now it is behind."

At the same university during the fall of 1981, a course entitled "Introduction to Problem Solving Using a Computer" was attended by 500 students. By spring of 1982, the enrollment jumped to 1,200 despite a warning in a student class evaluation book stating: "Nearly all the students complained of the large class size, poor facilities and/or inadequate terminal access." And in another similar course, a student was reported to write in: "I had the worst time fighting with 60 other students for the ONE terminal everyone had to use." Such comments are echoed on many campuses.

Estimates of the cost of new computer and engineering equipment vary considerably. Many of the differences are due to whether one is talking about teaching equipment (less expensive) or state-of-the-art research equipment (considerably more expensive). The Association of Independent Colleges estimated that in 1978, adequate teaching equipment at the bachelors level would cost approximately $1,500 per bachelor's degree per year. At the present number of 13,500 electrical engineering degrees, this comes to $20 million per year, or $130 million dollars worth of equipment with a useful life of 6.5 years. Estimates that include research equipment run much higher. Daniel Drucker, Dean of the College of Engineering at the University of Illinois at Champaign-Urbana, told a House of Representatives Committee in 1981 that the nation's engineering departments faced a $1 billion equipment deficiency and an equal amount for facilities and buildings.[18]

In its report on issues facing the engineering profession, the National Academy of Engineering addressed the issue of physical facilities such as classroom and laboratory buildings. "The physical facilities in which engineering colleges are housed are now about 30 years old. The federal government has not provided universities with funds for bricks and mortar since about the mid-1960s."[19] Through the late 1960s, federal funding was running at a rate of $120 million per year. By the end of the decade, the rate had dropped to $40 million.[20]

THE COST OF EQUIPMENT:

Fifteen Universities Add Up Their Needs

Expenditures Over the Last Four Years (1978-81) and Projected Needs Over the Next Three Years (1982-84) for Research Facilities and Special Research Equipment at 15 Universities*

(In thousands of dollars)

Department	New Construction		Modernization, Major Repair & Renovation		Special Research Equipment		Totals	
	78-81	82-84	78-81	82-84	78-81	82-84	78-81	82-84
Biological Sciences	$ 49,830	$ 45,503	$ 27,900	$ 27,038	$ 1,329	$ 4,660	$ 79,859	$ 77,201
Chemical Sciences	15,732	69,950	12,192	45,082	11,468	14,588	39,392	129,720
Earth Sciences	13,648	34,733	8,478	14,012	3,917	10,604	26,043	59,349
Engineering	13,963	107,012	24,052	76,094	27,058	33,222	65,073	216,328
Physics	5,392	54,546	12,125	20,179	5,695	22,590	23,212	97,315
Medical School	102,929	133,470	59,863	37,840	4,589	13,167	167,381	184,477
Total	$201,494	$445,214	$144,610	$220,245	$54,056	$98,931	$400,150	$764,390

* The 15 universities are University of California-Los Angeles, Cornell University, Washington University, Princeton University, Stanford University, Brown University, Northwestern University, University of Wisconsin, University of Florida, Pennsylvania State University, Massachusetts Institute of Technology, University of Utah, University of Maryland, Georgia Institute of Technology and University of Illinois. Figures may not reflect reports from all 15 universities.

Source: Association of American Universities

Budgetary cutbacks are a severe constraint on U.S. engineering departments, which have the will to expand but not the financial means. Salaries remain below competitive levels, student faculty ratios are far above normal standards, and laboratory equipment is overage and overused. For most of the country's engineering institutions, money and other forms of direct aid would seem to be the key to expanding capacity and increasing the output of EE/CS graduates.

But even if enough money were available — and in Chapter 7 we attempt to estimate the cost — there are serious doubts that the problem is just a question of money. The system of tenure also impedes a university's ability to respond quickly.

Student interests shift on a yearly basis; the faculty's changes on more like a thirty year basis. Also, the problem goes beyond higher education itself. A look at what is happening in American high schools, both in terms of quality and quantity, suggests that the shortage in human resources may get worse before it gets better.

High School Standards

High school standards are still falling and no bottom is in sight. The average quality of students graduating from high schools is at a fifteen-year low. This means that the pool of students entering college is not only smaller but less prepared than its predecessors. It also means that college level engineering departments will have to spend time and money on remediating what the high schools failed to impart.

In the high school class of 1980, 96 out of 100 graduating seniors had taken two or more years of English study, but only 66 percent had taken two or more years of mathematics. Only 51 percent of all high schools require more than one year of math for graduation. Achievement levels in math and science, tracked by the National Center for Educational Statistics (NCES), all showed significant declines for 9, 13, and seventeen-year-olds between 1969 and 1977. In physical sciences, seventeen-year-olds showed a 6 percent lower ability by 1977. This profile confirms a "general picture of modest decline in average science and mathematics achievement for the nation's youth from the mid-1960s to the late 1970s," according to NCES.[21]

College Board scores, however, show declines that are anything but modest for the nation's high school students who take the Scholastic Aptitudes Tests (SATs). They reveal a 22 percent drop in the number of top-scoring mathematics students: in 1972, a total of 93,868 students had mathematics scores of 650 or above versus 73,386 in 1980. The drop in verbal abilities was an astounding 46 percent. Average scores in math have dropped from 502 to 466 from 1963 to 1980, and verbal scores from 478 to 424 during the same period. Interestingly, the scores of the top 5 percent of the students have not declined, even though the overall average has.

Are these deteriorating conditions the fault of teachers? In the words of United States Department of Education Secretary, Terrell H. Bell, "teaching is not a profession pursued by the most academically talented and personable students on our college campuses. . . . Teacher education programs do not produce bright, knowledgeable and perceptive scholars."[22]

As teachers see it, the problem is far less of their own doing and more a result of budget cuts, drains by industry of talented teachers in science and mathematics, and public attitudes that seem to have turned away from education as a priority. "One would be hard put today to say that a school teacher is viewed as a role model or as a key figure in his or her community," says Dr. Max Sobel, president of the National Council of Teachers of Mathematics. "We need to restore the confidence of the public in the teacher."[23]

Dr. Paul Gray, President of MIT, points to the system of electives as part of the difficulty. Often, high school students are offered science and math as elective subjects. This is akin to a cafeteria where the student loads his or her tray with only the tastiest courses. Says Gray, "the teaching of science and mathematics at the secondary school level is a national scandal."[24]

A major part of the problem is the low salary paid to teachers. In the past two decades, this situation has reached the point where teachers now make less than garbage collectors in some cities. The whole attitude of the helping professions such as teaching, nursing, and others has changed. Previously teachers were underpaid and respected. Now everyone, regardless of status in society, wants to be paid well. Yet school districts are simply unwilling to pay teachers enough so that better qualified people enter the system. The entire educational infrastructure, not just science and engineering, is caught in this downward spiral. Most members of American society are going to have to learn to use computers in the future. Yet how can we expect them to do so with such inferior education in science and math relative to other countries such as Japan, Germany, France, and the Soviet Union? Even Britain, where engineering has a lower status than in the United States, is waking up to this issue.[25] The United States needs to develop a sense of urgency about the importance of schools to its future.

TWO CASES: Massachusetts and California

The Merrimack Valley Case: Five Percent Knew Algebra

Forming the border between Massachusetts and New Hampshire is the Merrimack River Valley, once an industrial area for New England shoes and textiles and now home to a high concentration of high technology employers. Northern Essex Community College (Massachusetts), a prime source of potential employees, is an active and modern school with an enrollment of 7,500 students.

Only 5 percent of the students typically know algebra when they enroll. In some of the valley's neighboring high schools, such as Lawrence High School, there is no math requirement for the granting of a diploma. Many students enter with low reading abilities as well, some rated as low as only third grade level.

Source: Dr. John Dimitry, president of Northern Essex Community College, Essex, Massachusetts.

High Schools in Silicon Valley: Low Interest in High Tech

"If Silicon Valley is any example, neither the schools themselves nor industry is taking major steps to strengthen existing [high school] programs or develop new ones that reflect . . . the shift in the nation's economy toward high technology.

"Moreover, a survey of the career plans of over 9,000 students in the ninth and eleventh grades . . . shows no significant interest in scientific or technical careers. Like students everywhere in the U.S., the most popular career choices were performing artist, doctor, pilot, lawyer, and professional athlete. Electrical engineer ranked 11th on the list and electronic technician 42nd.

"Education is in a dismantling mode," said one education official. . . The fact that $100,000 worth of electronics equipment sits idle in a high school classroom only several miles from some of the leading high technology companies is mute testimony to the gap between high school training and industrial needs. More ironic, the electronics instructor, recently laid off from high school in a round of budget cuts, now helps manage training programs for electronics technicians in a nearby semiconductor company."

Source: Dr. Elizabeth Useem, "Education and High Technology Industry: The Case of Silicon Valley," 1981.

Math teachers in high schools are leaving the profession faster than they are being rehired. The average salary paid a classroom teacher in 1980-81 was $17,678. Starting rates for a teacher with a bachelor's degree are closer to $10,000. The National Science Teachers Association in 1980 alone lost 1,000 of its 10,000 high school members. The lure to industry is attractive. In the words of two authors on the subject, "they can double their salaries just sitting at a desk as low level computer technicians."[26]

North Carolina found that in 1980, 45 percent of its math teachers were not formally certified to teach math. Virginia reported large gaps between needed math teachers and graduates ready to fill them. Nationally, a declining market for teachers due to enrollment cutbacks is leading the better qualified teaching candidates to look for jobs in other fields. For the math or science candidate, the high technology corporate employer is an obvious first choice. Dr. Max Sobel sees the country "as still able to turn out leaders to get us into space . . . but unable to train the mass of citizens who need math skills for a technology age."[27]

While evidence accumulates that a quiet crisis is brewing as secondary schools feel the lack of good teachers, the Reagan budget exacerbates it. Cuts in the Department of Education's budget of about 16 percent from fiscal year 1981 to fiscal year 1983 bring little solace to the teaching profession. One line item entitled "Professional Development" which included funds for teacher centers, a teacher corps, and precollege science teacher training was reduced from $10 million to zero in one year.

THE WASHINGTON STATE CASE: Where Have All the Teachers Gone?

In 1980, two professors at the University of Washington asked a bothersome question: "Where have all the teachers gone?" At first, the authors state, "we found that that picture was not at all clear. For example, while the National Educational Association was reporting a 28 and 31 percent surplus in mathematics and science . . . and a 79 percent overall secondary surplus, the National Center for Education Statistics had indicated an actual nationwide shortage of 1100 mathematics teachers and 400 science teachers. . . We decided to study the situation in our own state."

What they found was discomforting. For secondary school science teachers, in 1978-79 demand for 142 teachers was matched by a supply of only 86; and in mathematics only 54 new teachers were supplied versus a demand for 173. The authors predicted that the gap would grow more acute. "These results," they conclude, "suggest there will be a number of minimally qualified teachers filling science and mathematics positions in the 1980s."

Source: Roger G. Olstad and Jack L. Beal, "The Search for Teachers," *The Science Teacher*, April 1981, pp. 26–28.

While American high school performance in math and science declines, in Japan it remains consistently high – ironically, with the help of American research. According to the National Science Foundation, "there has been significant effort in recent years to revise and update the high school science teaching in Japan, and it is now heavily influenced by the U.S. Physical Science Study Committee (PSSC) and the Biological Science Curriculum Study (BSCS) materials."[28]

In the same study, the National Science Foundation summarized the findings of the International Project for the Evaluation of Education Achievement, led by the eminent Swedish educator, Torsten Husen. Japanese thirteen-year-olds ranked highest in mathematical achievement among the twelve nations compared – including the United States and a number of European countries.

Many observers feel that the problems in math, science, and engineering will be fixed by market forces within a four-to-five-year lag period. But the problem may no longer be so short term. As Steve Jobs, founder of Apple Computers suggests, we may now have a twenty year lag – the time between a child's first introduction to math and science and becoming a professional engineer or scientist.[29] Current high school (and elementary school) deficiencies in math and science can permanently cut off a large potential pool of children who might otherwise sustain interests in math and science. The problem is compounded if one considers the cultural bias that excludes women, early in life, from careers based on the sciences or mathematics.

Women and Minorities in Science and Engineering

"There is an enormous potential of talents available by bringing women into industry," says Congresswoman Margaret M. Heckler of Massachusetts.[30] Yet this potential, as well as that of minorities, is vastly underused. Between the recognition of what women could offer and their incorporation into the economic mainstream lies a long road. Only 8 out of a 100 bachelor's degrees in engineering were awarded to women in 1979. Less than half that number were awarded to women in mathematics. In contrast, almost all of the bachelor's degrees in library science were offered to women, as were the overwhelming majority in home economics, health professions, foreign languages, the fine and applied arts, and education.[31]

Changing demographics suggest changes in the role of women in science and engineering. Across the nation, the numbers of student-age college entrants are declining. In some educationally important states like Massachusetts, the numbers will decline by a striking 40 percent by 1991 – almost double the national rate. If the quality and quantity of enrollees in undergraduate programs are to remain high, a source of new entrants will have to be found. In engineering fields, women and minorities may be just such a supply.

The entry of women into the traditional male domains of mathematics, science, and engineering is not without immense problems. Without major changes they may be insurmountable. Generations of cultural perceptions that males are inherently more versatile in math and science are difficult to break; and despite repeated evidence that young girls and boys are equally adept, the culture tends to shunt girls out of a science career path early in their lives.

In *U.S. Woman Engineer*, the magazine of the Society of Women Engineers, Jewell Plummer Cobb describes the "filters" that progressively reduce the number of women able to pursue a career in science. Starting with early childhood when the boy gets a truck and the girl a doll, the young child identifies later with the predominantly female elementary school teacher who "most typically, but not always, expresses math anxiety at some point." By the time girls reach high school – the third filter – they are already predisposed to perform less well in math than their male counterparts. At home the father becomes the "math authority" while at school societal pressures convince the girl that poor performance in math "is due to a lack of ability rather than to a lack of effort. . . . And so as women proceed through life and through our educational systems, the filter gets finer and finer until few manage to move through them at all."[32]

It may be in college that the external and peer pressures are greatest. According to Ms. Cobb, the college environment does everything to discourage a woman from a career in the sciences. Evidence from the outside world seems to confirm this. Statistically, fewer than 3 percent of engineering jobs are filled by women, and the stimulus to change this pattern is still lacking.[33] "A female MIT student majoring in aeronautical engineering recounted that she had to fight all her life to retain her interest in aeronautics; her friends thought she was crazy, her mother told her she'd never find a husband, and her teachers warned her she'd never find a job."[34]

Graduate school aggravates the problem. Because the traditional expectation is that women will not complete their graduate career, faculty sponsorship tends to be diverted to male students. Additionally, at a time of funding shortages the male is favored even more as the competition for fewer grants and projects accelerates. This leads to the final filter: the senior scientist/adviser who will be the ultimate arbitrator of her acceptance or rejection for jobs, publishing, and general career development — a world in which the absence of senior women scientists or engineers makes the mature women scientist still subject to male authority.

But even at Stanford University, a model of higher education efforts to increase female participation in the science and engineering fields, difficulties abound. In 1980, one quarter of the undergraduate enrollees in engineering were women. This compares to a national average of one in ten. Yet despite the enviable record, unofficial university literature decries the fact that "the number of women full-professors [in science and engineering] can be counted on one hand; the number of women associate professors can be counted on two. A critical mass of women in science and engineering is yet to be formed."[35]

Judith Lemon, assistant director of the Stanford Instructional Television Network, coauthored a document on Stanford Women in Engineering. "Our purpose was to document who the women are who pursue engineering and science careers. It should serve," according to Lemon, "as a role model for others who think that women in science are abnormal in some way."[36] The document profiles forty-one women. Professional activities are juxtaposed against brief vignettes of personal interests. Prepared under the auspices of what the students have called W.I.S.E. (Women in Science and Engineering), the glossy, well-designed booklet reflects the organizational aims of W.I.S.E.: "To provide peer support and networking opportunities for Stanford's undergraduate and graduate women in sciences."

National statistics, however, point to a long journey ahead for women in science. In 1979, 57,593 men graduated with bachelor's degrees in engineering; 5,207 women received the same degree. Master's degrees were conferred on 14,558 males and only 952 females; doctoral degrees were awarded to 2,423 males and only 83 females – a telling reminder of the current biases in education.[37]

The situation for minorities is even worse than for women. Presently, only about 3 percent of engineering degrees are awarded to minorities, a figure which could be increased substantially.

Options

What are our choices for dealing with shortages in the technical workforce? To slow the growth of high technology employment to match the available supply of engineers? To increase the numbers of new engineering graduates? To increase the immigration of engineers from abroad? To increase the productivity of the existing engineering workforce? All of these possibilities will occur to some extent as the gap between supply and demand widens.

High technology represents a vital opportunity for substantial employment growth in the U.S. economy for the next ten to twenty years. Thus, irrespective of the expected demographic decline in the high school age population, we need to adopt a long term strategy to increase the nation's engineering educational capacity. From the statistics cited earlier in this chapter, it is apparent that at least a doubling of the capacity for certain disciplines will be required like electrical engineering and computer science. But even with a concerted effort, it will take three to five years before the size of the existing engineering workforce can be increased by more new graduates. Therefore, in the near term, the best alternative is to increase the productivity and retention of our existing pool of experienced engineering talent.

If we could increase the productivity of the approximately 200,000 electrical engineers now working in industry by ten percent, that would be equivalent to 20,000 new engineering graduates — or perhaps more, given the experience factor. With new CAD/CAM equipment and other powerful electronic engineering aids now available, this should not be a difficult goal to achieve.

As we shall argue in a later chapter, there is also an opportunity to increase the retention of engineers and extend their professional working life. Presently, too many maturing engineers leave their profession for better-paying and more influential jobs in management. What is needed is to elevate the prestige and professional status of engineering to a level where more talented people will not only enter but will stay in this field to cope with increasingly complex problems.

Finally, we cannot overlook immigration as a source of technical talent. U.S. authorities make it difficult for foreign students to stay and work in America. For experienced engineers, it is almost impossible to obtain work permits or permanent visas, even for those in areas critically short in supply. Congress is actively considering more restrictive legislation that would require foreign nationals to return home for two years after receiving their degrees before they can be eligible for employment. This is another of many indications how one set of policies conflicts with others needed by the country, and how quickly the heritage that has made America a great nation is forgotten.

Faculty shortages, capacity problems, declining high school standards, and an underuse of potential sources of talented scientists and engineers are matters of grave concern. Viewed from a distant vantage point, they indicate a decade or more of failure by policymakers to construct a cohesive and coherent long-term education and training strategy. Equally disconcerting is that in 1980, national policymaking exacerbated this trend by de-emphasizing education as a central national policy concern and by disbanding any serious long-term effort to retrain and re-educate workers displaced by structural changes in the job market. Instead, defense has become the centerpiece of national policy. This even further complicates the human resource dilemma, as we shall discuss in the next chapter.

Chapter 5

DEFENSE PRIORITIES

NATIONAL DEFENSE is intimately interwoven into the day-to-day affairs of the nation. Yet if one listens to the current administration, one finds little attempt to link a projected five-year, $1,600 billion military budget to the health of the domestic economy or to the global competitiveness of U.S. industry. The national priorities are established as if defense were detached from other considerations.

This chapter focuses on two defense-related questions. The first is whether the race to create the fastest computer chip will help or hurt high tech industry. Both private industry and the federal government are engaged in building faster computer chips for the 1990s. The former is doing it to stay competitive in a global marketplace; the latter is doing it because a new generation of electronic weaponry is dependent on ever faster and more complex computer controls. The problem is that the two priorities not only pull in two different technological directions but they conflict in vying for a limited number of highly qualified engineers. Thus the Department of Defense may be working at cross purposes with industry by stealing away scarce manpower for its new fast-track R&D programs.

The second question is the broader guns versus butter issue that has been debated since President Dwight D. Eisenhower's military-industrial-complex warnings. At what point is our national security better served by a stronger economy than by more outlays for military hardware? Ronald Reagan's $200 to $300 billion annual defense outlays cannot be sustained without serious effects on the domestic economy. One effect is the administration's decision to reduce the nation's investment in education at the very time when more and better qualified engineers, scientists, and technicians are needed by the private sector *as well as* by a military complex highly dependent on high technology.

One of the striking differences between the present bulge of government defense spending and the defense and space thrusts of the 1950s and 1960s is that, in the earlier period, the creation of an enlarged and up-to-date technical workforce was seen as an integral part of the program. Early investments in government R&D made a significant contribution to subsequent commercial development, long after defense and space priorities subsided. This time, however, the policy is to dip into the existing technical pool to support defense priorities. This diverts manpower resources from commercial growth without a lasting contribution to long-term economic development. In the long run, the build-up of military spending may be not so much an issue of budget deficits as it is a failure to match military spending to a build up of the nation's technical manpower base.

The Fastest Chip

A SIGN OF THE TIMES: Spring Course Offering at MIT
Submicrometer Structures Technology

Can 10^{12} bits of information be stored on 1 cm^2? Can we fabricate surface structures with tolerances $\sim$ 10 Å (Comparable to the size of organic molecules)? The technology for doing so may be in hand within a few years, leading to revolutionary applications in electronics, optics, materials science and organic chemistry. This course will survey methods of fabricating and analyzing articifical structures having submicrometer dimensions, and discuss selected applications.

After a lapse of nearly ten years, the federal government is again playing a major role in the development of computers and electronics. An example of one of the target technologies is described in the MIT course offering above — "submicrometer structures technology," or, in layman's terms, the new technology of making faster chips. The new computer and memory chips of the 1980s and 1990s are known in the industry as VLSI and VHSIC. Private corporations have invested heavily in the development of Very Large Scale Integrated Circuits (VLSI). These are becoming the new standard components in the latest generations of computer and electronics hardware. At the same time, the Department of Defense (DoD) has begun to produce new chips to control the next generation of missiles, radar, satellites, and aircraft. These are known as Very High Speed Integrated Circuits (VHSIC). The VHSIC program for the fastest chip encapsulates many of the issues in the debate on whether greater defense expenditures are in the long run beneficial or detrimental to the nation's economy and high technology industry.

The VHSIC program, which is funded out of what the DoD terms "systems deployment budgets" rather than out of *pure* research funds that exclude development, will cost $300 to $500 million between 1979 and 1986.[1] While not staggering in terms of the overall defense budget, these amounts represent a significant increase in the department's basic annual pure research, estimated at $320 million for 1982 alone after an inflation-adjusted increase of 17 percent over 1981.[2]

What is so special and valuable about VHSIC chips to warrant these expenditures? The primary answer is speed. Thumbnail in size, by 1990 these devices will perform at speeds unimaginable to the layman — from two to twelve billion additions or multiplications per second. This represents an increase anywhere from 10 to 100 times faster than today's fastest chips. The requirement for increased speed depends on the application: for radar, 50 times faster; for weapons targeting, 100 times faster; for electronic warfare, 50 to 200 times faster.

While high speed is most emphasized, the fastest chip is actually a synthesis of many different specifications. It is to be smaller by a factor of ten, highly resistant to radiation, cheap enough to be "throwaway," and light enough to be portable. For the Army's battlefield information distribution system, the weight factor will be reduced from trucksize to backpack size. For the Air Force's advanced electronic warfare surveillance system, what in 1960 would have required more than 270,000 separate circuits weighing more than a two and a half ton truck will by 1990 be compressed into 6,000 chips with a weight of only 180 pounds.

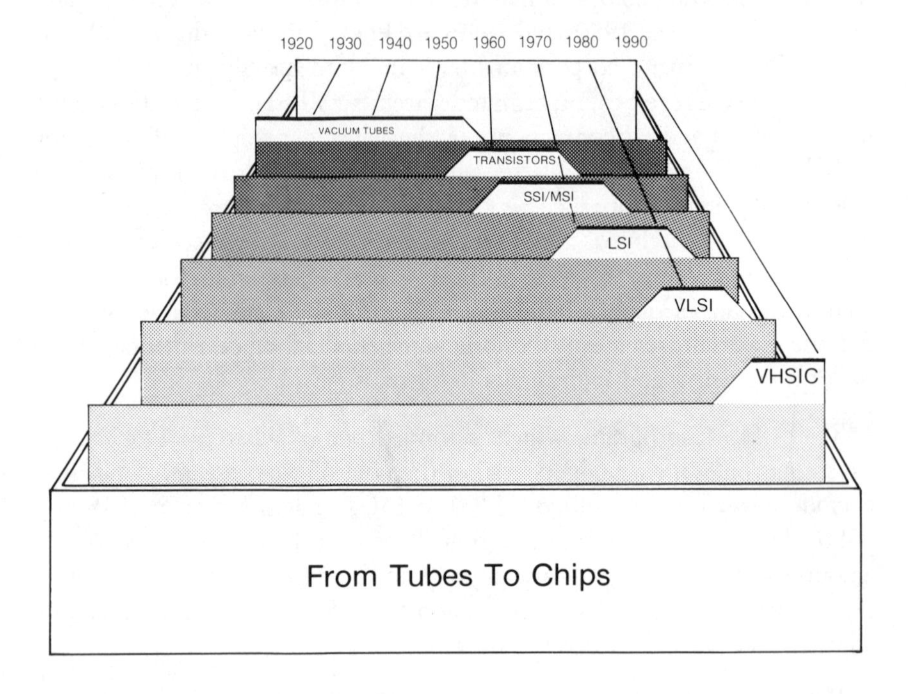

This technology is the latest culmination of an electronics revolution that started with vacuum tubes in the 1920s. After World War II, the American invention of the germanium transistor was quickly capitalized on by the Japanese, who successfully applied miniaturization technology to capture large segments of the consumer electronics market. The United States, relinquishing the transitor based market, moved rapidly into a new generation of silicon miniaturization prodded by the military's needs for high performance computer and communications technology.

Between 1950 and 1970, federal contracts helped bankroll developments that would revolutionize electronic technology. During this period the government spent $900 million (1965 dollars) in semiconductor research and development. Companies such as IBM and Control Data were early beneficiaries of major defense and NASA contracts. During the late 1950s, almost six out of ten revenue dollars at IBM were government related. These subsidies and contracts were instrumental in catapulting the computer industry to world dominance. In addition, the government's procurement of equipment channeled immense amounts of working capital to the high technology industry.

A key turning point in the growth of integrated circuit (IC) development was a combined decision in 1962 by the National Aeronautics and Space Administration (NASA) and the Air Force to use ICs in the Apollo program and in the electronic guidance of the Minuteman Intercontinental Ballistic Missile (ICBM). The initial Apollo orders, most of which were placed with the Fairchild Corporation, required delivery of 200,000 circuits. Texas Instruments and Westinghouse produced circuits for the Minuteman program. Soon thereafter Motorola and Signetics became suppliers.

During the 1970s, private sector markets surged upward with the coming of mini and micro computers. This quickly displaced the federal government as a primary buyer of new electronic technologies. Its share of purchases declined rapidly to a current rate of about 10 percent of the U.S. production of semiconductors and 33 percent of all imported ones. By 1975, the private market had outpaced the military as a buyer of electronics technology end-products. Companies such as Digital Equipment Corporation, Wang Laboratories, Data General, and others Computer have since risen into the Fortune 500 ranks as hardware suppliers to the commercial marketplace. The number of chips produced by commercial manufacturers since then are logged in billions – an amount that challenges the cumulative appetite Americans have had for McDonald's hamburgers, at a unit price that is roughly equivalent.

By 1980, military sales of computer and electronics equipment represented small portions of the total sales of the major firms. For example, out of the top 100 defense contractors, IBM ranked only twenty-eighth with $500 million or 2 percent of total sales going to defense; Control Data was seventy-fourth with $140 million or 5 percent of total sales to the military sector.[3] But by 1981, defense expenditures were rising again and dramatically; Ronald Reagan added momentum to a pattern already under way under President Jimmy Carter. Introduction of the VHSIC program under the latter's administration was a distinct sign that the DoD would once again make the federal government a major force in the development of state-of-the-art electronic technology.

The Scarce Skills Problem

In Chapter 4, we described the shortages of two of the most important skills needed for VHSIC and VLSI work – process engineering and semiconductor design. These were shown to be in far greater demand than supply: a current nonmilitary demand of 4,500 individuals against a total supply of only 2,450.

Contractors and subcontractors alike expect to switch their best engineering talent to the VHSIC project because it requires the most advanced skills. This might not in itself represent much of a problem, except that there is some question as to the commercial impact of military research. The most immediate impact will be to leave potentially large vacuums in nonmilitary research efforts, especially at highly experienced scientific levels. Yet it is precisely those skilled, mature engineers who are most difficult to find and that industry most needs.

The situation is somewhat different at the universities. The third stage of VHSIC work has been classified. This is a serious turn of events and if extended too far could lead to retarding university research in microelectronics. As a result of the new secrecy moves, Cornell University has withdrawn from the VHSIC research program, and other leading universities will likely make a similar move.

At a recent conference on VHSIC, Ivan Saddler from Motorola summed up the situation in process and design engineering by contrasting their employment status to that of professional athletes. Relative to sales, there are ten times as many professional athletes as process and design engineers. Yet athletes have free agent drafts, five and six figure salaries, and multi-year contracts! If there are shortages of scientists and engineers in 1980, we can guess what could happen to the semiconductor industry and salary rates by the late 1980s with programs like VHSIC.[4]

While many people are willing to venture guesses, no single authority is willing, or perhaps able, to predict what some of the more pronounced effects of the VHSIC program might be. There is, however, general agreement that rising salaries will be the first whistle from the pressure cooker. If precedents are any guide, the DoD's tendency toward cost-overruns will inadvertently encourage industry to bid up prices in order to hire the right people to meet the project's stringent goals. This will draw the limited supply of skilled workers toward the higher paid defense contracts, which in turn could have a negative impact on the commercial sector of the industry.

Some companies, like National Semiconductor and Signetics, are confident that as much as 85 percent of the VHSIC tasks are compatible with their commercial goals. If this prediction proves accurate, then the initial negative impact will turn out to be positive in the long run. Others, however, such as Intel and RCA, spurned the contractual opportunities, feeling they would divert resources and capital to purely military purposes and weaken their marketplace competitiveness.

How Much Fallout?

How much of the VHSIC research will actually have commercial application? Will it strengthen or weaken the United States vis-a-vis Japan, where military expenditures are miniscule?

Much of VHSIC is designed to meet specific military needs. The chips are part of larger operating systems such as radar control for tactical fighter airplanes or battlefield information systems for field commanders. The specifications call for extremely high radiation-hardness and an ability to withstand temperatures from −55 degrees to +125 degrees centigrade (−67 and +257 F). To Erich Bloch, a corporate vice president of IBM (see Chapter 6), features such as these "can only serve to side-track commercial efforts by setting R&D standards that are not compatible with industry's needs."[5] Unique military specifications may make it difficult to separate commercially applicable chips from highly integrated military hardware. Or, as one observer put it, it's hard to sell sophisticated radar equipment at the local Sears outlet.

Even if military research turns out to have relatively little direct application, there are bound to be some positive indirect effects. One lies in advanced processes. Signetics, for example, sees major benefits coming from improvements in materials, lithographic patterning, electron beam writing, and other manufacturing processes. Another indirect effect is the secondary support of research laboratories. The benefit of defense contracts during the 1950s and 1960s was in subsidizing an entire generation of skilled individuals who could later apply their knowledge in the purely commercial marketplace.

Whether renewed emphasis on defense research and development will sustain university research centers and the specialized teams of professional scientists who will in time flow into commercial areas is an unanswered question. In the shortrun, the prospects do not favor the commercial sector. The immediate problem of scarce skills seems likely to outweigh any hope for marketable commercialization of defense research. In any event, this example of VLSI versus VHSIC illustrates that little thought is being given in national policy to the needs of the emerging high tech industry and how to superimpose huge new government demand in a way that the long-term growth of this industry will be aided rather than impaired.

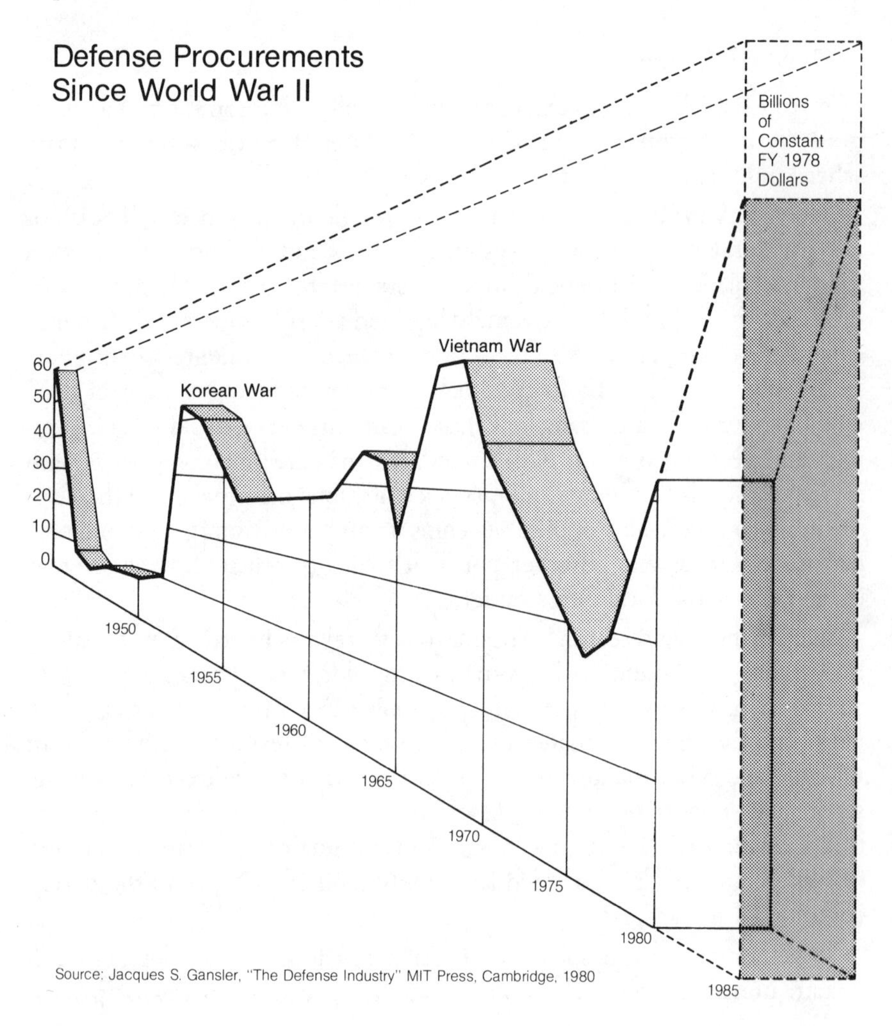

Source: Jacques S. Gansler, "The Defense Industry" MIT Press, Cambridge, 1980

Boom or Bust?

There is a larger context to the $1,600 billion military expenditure question. A fear associated with a surge in defense spending is of the "boom and bust" cycle that accompanies it. The boom comes at the expense of other priorities; the bust leaves whole sectors of the economy high and dry. This problem was acknowledged in 1979 by the General Accounting Office (GAO) in a report submitted to the Office of Science and Technology Policy. Addressing the problem of engineering and science shortages, it recognized that "government programs and policies greatly influenced the boom and bust cycles of science and engineering labor markets over the past several decades." It went on to observe that "shifts in domestic priorities caused large-scale fluctuations in demand for manpower, including highly trained manpower." The GAO concluded that the "inherent flexibility of the economy can not always be relied on to assure a balance between supply and demand."[6]

Defense procurement trends are a clear indicator of the ups and downs that can disrupt the normal marketplace for labor, much of it highly skilled scientific and engineering manpower. In February 1982, the widely read Rosen Electronics Newsletter talked about Reagan "procurement outlays going out of sight."[7] While the defense budget was increasing at an 18 percent annual rate, the research, development, testing, and evaluation budget was forecast to rise 21.3 percent in 1983, and the hardware procurement by 33.4 percent. According to the newsletter, "The procurement requests are dramatic, indeed."[18]

The employment effect is already visible in 1981 – especially in the aerospace industry, which is no stranger to the peaks and valleys of government-induced employment. Scientist and engineer employment rose in 1981 to 220,000 or about 12,000 more jobs than in the prior year. The same is forecast for 1982 and beyond to match anticipated increases in defense expenditures.

Bottlenecks are envisioned. Speaking before a congressional committee in October 1981, General Robert T. Marsh, Commander of Air Force Systems Command stated:

> President Reagan's defense budget will create an immediate demand for even more technically qualified people in programs such as the Long Range Combat Aircraft. Thus, I see my problem about to get worse. But this is not new, . . . the engineer shortage has been a fact of life for [my command] for more than five years. . . The

> Air Force as a whole is nearly 1,100 engineers short of our minimum needs. And Systems Command, as the primary user of this critical Air Force resource, is short over 500 military engineers — or ten percent. This shortage is particularly acute in the electrical, astronautical, and aeronautical engineering disciplines. . . . Like industry, we have recruited aggressively.[9]

In the civil aviation industry, because of a recessionary fall in domestic demand for aircraft, the military orders come as a welcome offset. But there are signs that the increase may be too large, and that what began as an offset may be a "strip out," either from other tasks or from other companies. For example, Buz Hello, the chief of the aircraft group at Rockwell International Corporation, the big contract winner on the B-1 bomber, has "identified 5,500 Rockwell professionals he could strip out from other assignments . . . while he would recruit 11,000 more from outside the company."[10]

In contrast to aerospace, the situation may be the reverse in the commercial electronics and computer industry, where there is no room for an offset on the "up" side. In other words, rising defense expenditures could further exacerbate the present shortages since they come at a time of strong civilian demand and intensive foreign competition. The problem is how to control the boom at a time of limited manpower supply in critically needed fields.

Not only is there concern about controlling the boom part of the cycle, but there are worries also about the bust. What happens when the high flying defense budget peters out? What happens after the economy and educational systems have geared up, programs such as VHSIC have run their course, and large numbers of computer scientists and electrical engineers are no longer needed by the defense department? This is just what happened in the late 1960s when large numbers of aerospace engineers were let go after the NASA program and other military programs were scaled down. Fortunately, the answer may be found in present industry trends. Continued growth of the electronics technologies aimed at commercial markets may offset the bust side of the military expenditure cycle. But part of that bullish growth expectation is based on the United States being able to retain its international technological advantage — the critical edge that is being squeezed by the boom in defense expenditures.

At present, the greater of two evils is less the danger of bust than the potential of overheating the boom. Part of what is needed is for the

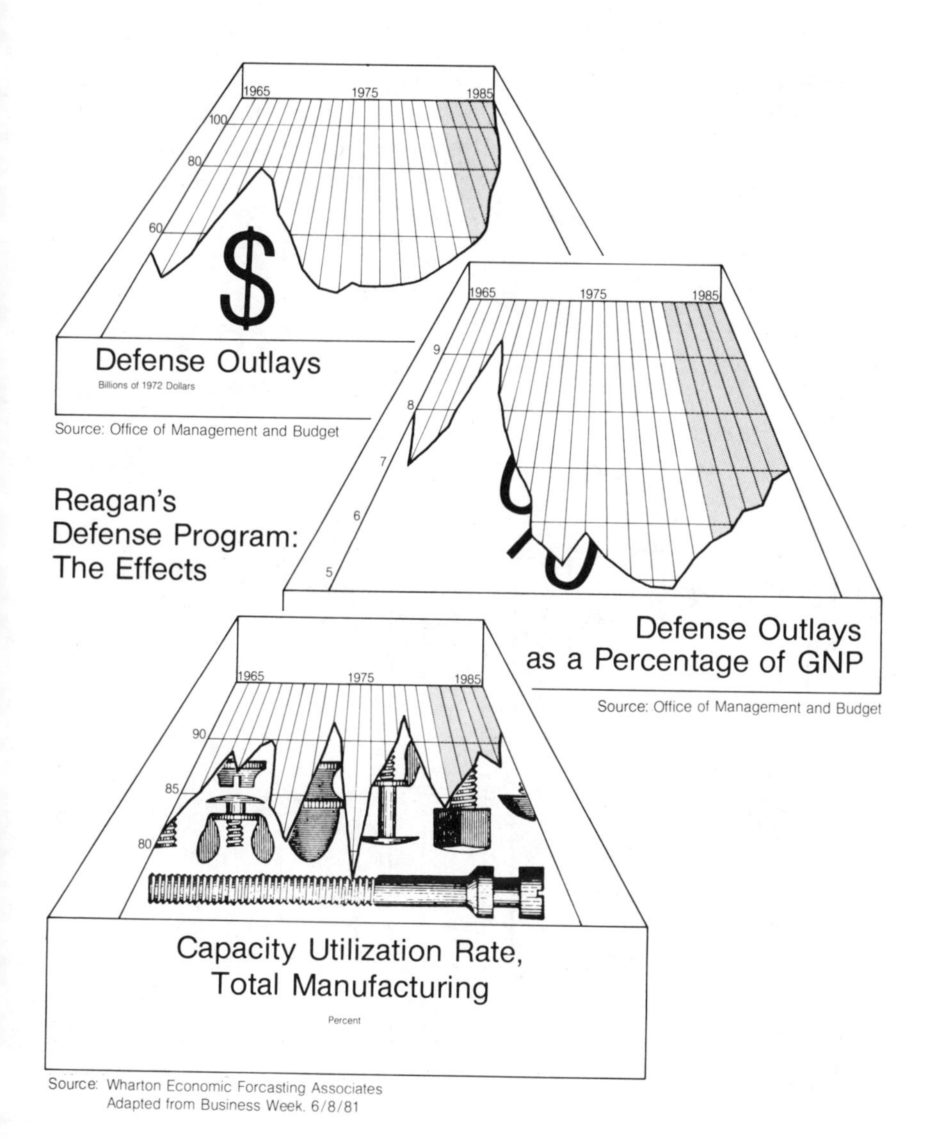

Department of Defense to contribute in equal measure to the education and training of those it will almost certainly hire in high demand areas. Otherwise private industry will suffer from strip out and have to pay the retraining offset price when defense programs trickle out.

Guns Versus Butter

President Reagan's $1,600 billion spending program on defense over a five-year period, brought to life on a tide of favorable public support, quickly turned sour. Part of the reason is that this estimate may be only

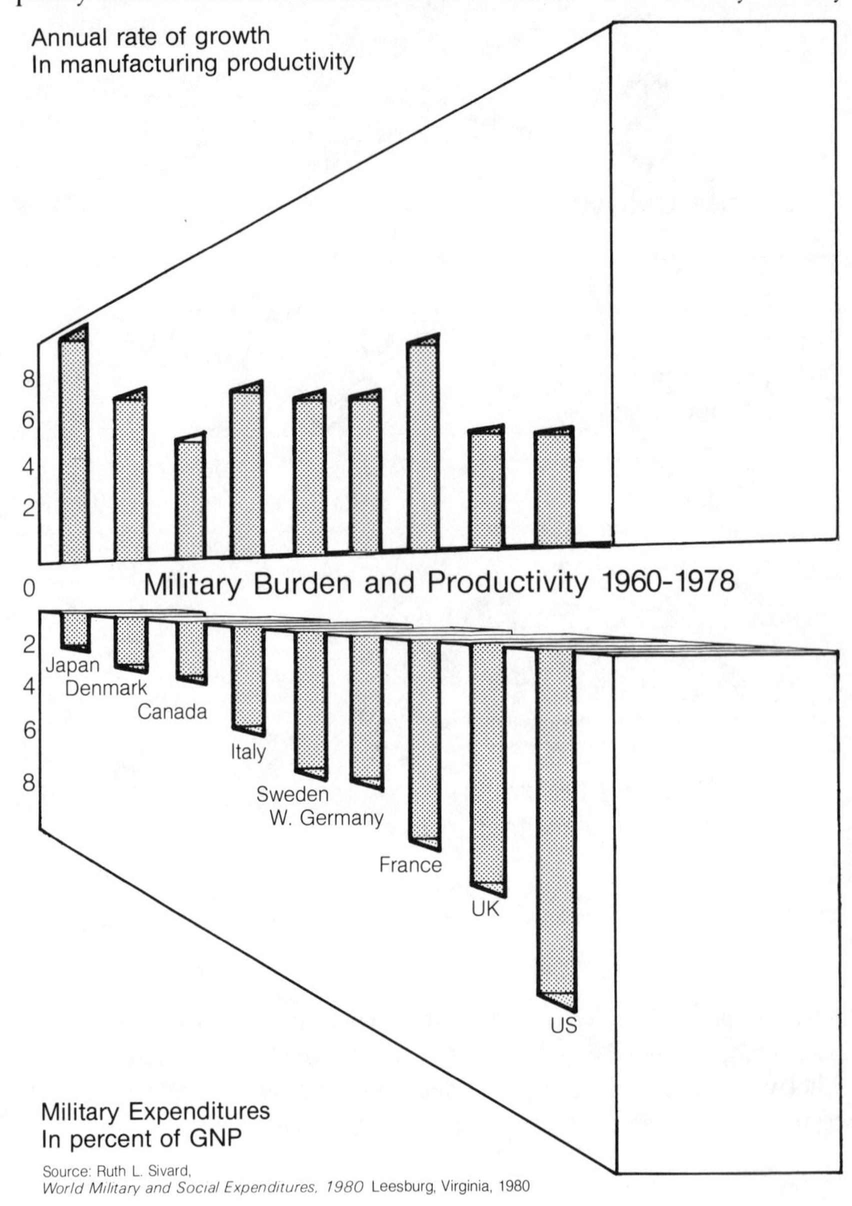

Source: Ruth L. Sivard, *World Military and Social Expenditures, 1980* Leesburg, Virginia, 1980

part of a far larger iceberg. According to the *New York Times*, Pentagon officials in March 1982 already foresaw a need for another $750 billion to meet program requirements specified by the Reagan administration.[11] But the principal reason for a vocal opposition gaining currency is that the sacrifices from the rest of the economy were becoming evident and controversial. Not the least is a fear that unattended social programs will breed a groundswell of explosive reactions – much as during the volatile urban crisis years of the mid to late 1960s.

Despite growing opposition, supporters of the big-bucks approach to national security are steadfast. Dr. David L. Blond, a senior economist at the DoD, wrote in 1980 under the title "Guns Versus Butter": "The macroeconomic models indicate clearly that the economy contains sufficient excess capacity, at present [1979], to accommodate higher levels of defense spending."[12] The controversial belief that American riches are sufficient to sustain a war economy without inhibiting its domestic appetite is a matter of fact to Blond.

This view is not shared by F. Gerard Adams, an economist at Wharton, who says: "The reality of the impact of the planned increase in defense spending could be worse than any macro models predict."[13]

What is the effect of sustained defense increases on the nation's productivity and on inflation? A comparison of several countries with varying proportions of defense expenditures in relation to their gross national product suggests that a guns or butter strategy is an either/or option; together they put an unbearable strain on a domestic economy.

This view was confirmed by five noted economists* who met on October 27, 1980, at the invitation of the Department of Defense to discuss the guns versus butter issue. Jimmy Carter was still president and it was not yet evident that Ronald Reagan would succeed him. It did not quite matter, because the subject would have direct bearing on the policies of either.

Among the questions asked of the five consultants was the possible effect of increased defense expenditures to annual rates of 10 percent above inflation (Reagan would soon settle on a real annual increase of 7

* Mr. Gary Ciminero, Vice President, Merrill-Lynch Economics
Dr. Otto Eckstein, President, Data Resources, Inc.
Dr. Michael Evans, President, Evans Econometrics
Dr. Lawrence Klein, Wharton Econometric Forecasting Associates
Dr. Leon Taub, Vice President, Chase Econometrics

percent). After a few weeks of consideration their conclusions were summarized in a memorandum to the Secretary of Defense:

> All participants agreed that to maintain economic stability, the 10% increase must be compensated by *reductions* in non-defense spending or by higher taxes. . . . Second, . . . higher levels of [uncompensated] defense spending if they come at the expense of new private capital investments designed to increase productivity will be *inflationary* in the long run. Third, the 10% rate . . . is probably the maximum sustainable growth rate. *Bottlenecks* will occur in some industries despite the modest impact on the full economy. Fourth, the increase in real output may lead to a widening of the *trade deficit* and a further depreciation of the U.S. dollar. As a result, the inflation rate may be 1% higher than forecast. A beneficiary then of the increased U.S. investment in defense would be Europe and Japan as well as the Third World.[14]

The question of guns versus butter in the 1980s is intimately tied up with supply side economics. Faith in the supply side buttresses the case for increasing both guns and butter at the same time. But what if supply side economics fails to work as promised? Lester Thurow asked that question, and the answers are shown in the graph below. The most striking aspect of the problem is the immense impact foreseen in the U.S. deficit.

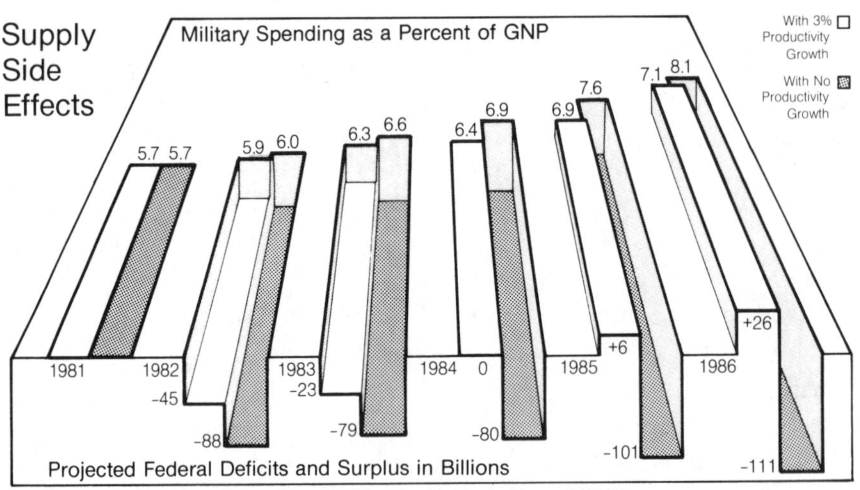

Source: New York Times; Beware Reagan's Military Spending; Lester Thurow, 5/31/81

The gamble is clear. Budgetary expectations of DoD officials bank on supply side successes. Dr. Blond argues that the United States can absorb increased defense expenditures without undue negative effects on the economy. In his article "Guns Versus Butter," Blond insists:

> The US technological base that exists today will allow our defense program to grow while civilian requirements are adequately met. . . . I have heard it once and I have heard it a hundred times, expenditures on national defense are highly inflationary because they take goods and/or productive resources from the private economy without adding products that individuals may consume directly. . . . This is an argument more out of the 16th century than the 20th century. It is an argument dependent upon a simplified, free market, almost agrarian society; and it does not fit today's differentiated, complicated, quasi-free market, advanced technological society. . . . I will attempt to defeat the myth [that guns preempt butter] using the weapons economists have developed to dispose of imposters – macroeconomic models and logical deductions. I know, however, that in the end the myth will, like all illusions, recur and defeat me."[15]

The Bottom Line

In the long run, the ultimate success or failure of VHSIC and the DoD's attempts to produce the fastest chip are linked to the success of supply side economics and whether or not we can generate enough wealth to afford both guns and butter. But in the short run, VHSIC will require reassigning scientific and technical talent from sources yet to be specified. The implicit message from supplysiders and defense economist Blond is that it is unlikely for the Department of Defense to look beyond the open market as a supplier of its manpower. In other words, the department will probably not undertake any additional commitments to education as part of the VHSIC program or any other aspect of its needs.

This apparent cold shoulder to education even in the hard sciences is a new twist for the American government. In earlier years, especially in the post-Sputnik era, education was seen to be an inherent part of defense policy – witness the series of National Defense Education Acts of the early 1960s. Had America in the 1960s been faced with the problems of the 1980s – intense international competition, human resource shortages, and rising defense expenditures – a logical and compelling response would have been a major new effort in education. There would have been calls for more science and engineering education, foreign language training, computer sciences, and greater efforts at reading and writing skills. But this time around, national defense policy does not take into account the needs of the emerging knowledge-intensive industries.

While the federal government and the DoD move to downplay education, private industry and a few state governments have another idea. With little fanfare but great effectiveness, some of America's largest high tech corporations are responding to these national challenges in innovative ways. New alliances are being formed, and new partnerships are emerging between corporations and universities and occasionally some three-way arrangements of industry, academia, and state government. In the next chapter, we describe six such emerging partnerships, and profile some of the individuals who lead them.

Chapter 6

RESPONDING TO THE CHALLENGE

THREATS FROM Japanese competitors, declining conditions in American engineering departments, and a federal policy heavily skewed toward traditional capital investments and defense priorities together present a formidable drain on the growth potential and competitiveness of the American electronics sector. The potential escalation of the threat, now more clearly evident than ever, is unleashing the creative energy of many of the industry's corporate leaders.

Similar reactions are visible from academia, where university presidents are acting more vigorously on behalf of their beleaguered institutional budgets and embattled engineering departments by seeking out new cooperative relationships with industry. So, too, for some state governments such as North Carolina, California, Massachusetts, and a few others that understand the need for far-sighted economic growth policies based on the fortunes of new industries. The imperative for a response to conditions that challenge the health of the U.S. economy is leading to new partnerships and new experiments especially between industry and academia, and occasionally with state governments.

New Alliances

The high tech industry has not been idle on the education front. An example: at a "Route 128" hotel conference room on a wintry Massachusetts day in 1982, a new coalition is in the making. The meeting begins at 7:30 A.M. on a Wednesday morning. Two hundred and fifty corporate executives and presidents of colleges and universities are gathered to review new ways and means of working together. For six hours they sit through speeches and commentary sprinkled with increasingly familiar phrases: "engineering education is at saturation," "remiss not to redress the shortfall," "twenty years of neglect," "a potential for disaster," "long-term partnerships," "lifelong cooperative education," "we've lost national conviction," "federal proposals are devastating," "strategic leadership needed," "the 2 percent solution." The speakers include a roster of computer company founders from Data General, Digital Equipment Corporation, Wang Laboratories, and Analog Devices. Their academic counterparts are presidents of major universities – MIT, the University of Massachusetts, Northeastern University, and others.

Sponsored by the Massachusetts High Technology Council and the American Electronics Association, the conference resulted in a good deal of press publicity for a resolve by many high technology companies to contribute 2 percent or more of their R&D budgets to help support higher education. But what one could not also fail to see was the political and social significance of the meeting. New alliances were being forged in ways that would have been unheard of in earlier decades. Whether they would be effective or lasting was not the immediate concern. The point was, it was beginning.

State governments, while a step behind industry in all but a few cases, are also showing signs of new activity. Late in February of 1982, the nation's governors met for their winter gathering in Washington, D.C. Central to their agenda was a review of America's high technology future. The governors' meeting bore a familiar resemblance to the conference of Massachusetts presidents less than two weeks earlier. Governors, industry leaders, academics, federal agency representatives, and various elected officials listened in a large and crowded hotel conference room to the familiar litany. Dr. Frank Press, president of the National Academy of Sciences: "[economic] growth will go where education is strongest and technology transfer occurs. . .; the crisis of pre-college education needs presidential leadership. . .; universities are in trouble." Senator John Glenn of Ohio: "We're breaking up R&D teams that took years and years to build." Congressman Timothy Wirth of Colorado: "For the first time we are producing a generation of kids less literate than their parents." Steve Jobs, Chairman of Apple Computer: "high school education is a disaster, we have to get computer equipment to them. . .; this is a knowledge-intensive economy and we aren't teaching our kids accordingly." Congressman Ed Markey from Massachusetts: "The Reagan budget cuts are synonymous to a book-burning."

If the words seemed strong, perhaps it was because of the politically charged aura of the meeting – potential new alliances in the making. This was expressed in a jointly signed statement by the cochairmen of the Task Force on Technological Innovation, Governor Edmund G. (Jerry) Brown of California and Governor William G. Milliken of Michigan. "Federal budget cuts have reduced the range of federally funded tools available to encourage increased investment in industries that will utilize technological advances. This has created an increased demand for state action, raising fundamental questions of the appropriate relationship between states and this industrial sector." And they added: "Industry itself should evaluate and discuss with governors the most productive state-industry relationship."[1] As an ironic footnote, the meeting was extensively taped by a Japanese television crew.

States like North Carolina, California, and Massachusetts are leaders in not only perceiving but actually implementing economic strategies founded on the projected long-term growth of emerging industries such as electronics and biogenetics. A number of other states such as Colorado, Arizona, Texas, Michigan, and Minnesota are forging similar strategies. These efforts mark the beginning of a political trend that is bringing many new (and old) names into national and regional currency; Governors Hunt of North Carolina and the already well-known Brown of California are headlined as far-seeing spokesmen for knowledge-intensive economies of the coming decades; and in Washington, D.C., congressmen like George Brown from California, Timothy Wirth from Colorado, and James Shannon from Massachusetts are identified with new legislative initiatives to resolve pertinent national high technology issues.

A new mood of action is discernible. And with it comes a renewed concern for the American future. The public press is replete with cases of corporate efforts to initiate university links; with state governments recognizing a new responsibility toward education and its relationship to growth; with academia redefining its mandate in recognition of diminished federal support. But rather than attempting a lengthy list of such activities, the authors focus on six examples that illustrate the search for new links among industry, university, and government.

The first focuses on a historic decision by twenty semiconductor and computer corporations to pool research budgets into a super fund in response to Japanese competition. The large size of the new fund and the decision of the group to focus its efforts through only a few select research centers counters a long tradition of research dependence on federal funds shared among a wider number of institutions.

A second example reviews the efforts of a University of Minnesota professor to initiate a working partnership with major corporations based in Minneapolis-St. Paul. Important aims were to establish a microelectronics center and to leverage corporate grants into programs that would increase the number of engineering students as well as link to industry experience.

A third example, in Palo Alto, California, provides a formula for industry initiative in creating a new research center, administered by Stanford University, but drawing its major operating funds from federal contracts.

A fourth example focuses on an ambitious response to a widespread need for support of engineering education across the country. The American Electronics Association created a foundation that could channel corporate contributions to engineering departments for salaries, scholarships, and computer-based equipment. This approach raises the basic question of whether volunteerism can not only meet immediate educational needs but also sustain the commitment over an extended period of time.

A fifth reviews the founding of a new academic institution, the Wang Institute of Graduate Studies. It represents a case of enlightened philanthropy by Dr. An Wang, whose belief that traditional academia would be adverse to radically new curricular departures prompted him to finance the new institute.

The sixth looks at one state, North Carolina, as a model of a constructive and visionary partnership led by state government and incorporating education and industry. This case raises two questions. Can state governments, independent of federal policy, build a national strategy? Does the North Carolina example offer a working model of what might be a federal role in meeting the challenges affecting American high technology?

An interesting feature of these six examples is that they reflect an entrepreneurial spirit and ingenuity of individual leaders. Their personalities and working philosophies help to explain the paths they took to effect change and to create new institutional ties. They remind us that the pioneering efforts and imagination of one person can indeed have far reaching effects. A glimpse into who these people are is as important as the description of the programs they initiated.

These include *Erich Bloch*, chairman of the Semiconductor Research Cooperative, and a vice president of IBM. *Dr. Robert M. Hexter*, a professor of chemistry who, with Control Data and others, initiated the Center for Microelectronics and Information Services at the University of Minnesota. *John Young*, president of Hewlett-Packard and key fundraiser behind Stanford University's Center for Integrated Sciences. *William Perry*, former DoD assistant secretary of research, who foresaw the need for an Electronics Education Foundation. *Dr. An Wang*, founder and president of the Wang Institute of Graduate Studies and founder of Wang Laboratories, and *Governor James B. Hunt, Jr.*, who successfully had $24 million appropriated for a North Carolina Microelectronics Center located in the Research Triangle Park near Raleigh-Durham.

Together these men and their initiatives symbolize the makings of a potentially formidable response to Japan, France, and other nations. They personify a uniquely American advantage – pragmatism and flexibility in creating new alliances and institutions to meet future needs. From a global perspective, these initiatives represent a test of strength between the diffuse tactical power of many American corporations, institutions, and individuals and the concentrated strategic will of governments like Japan and France. What do the emerging pieces of an American response look like?

THE SEMICONDUCTOR RESEARCH COOPERATIVE (SRC) (Semiconductor Industry Association)

SRC, a consortium of some twenty corporations, expects to channel eventually about $50 million per year into three to six selected American universities. The purpose is to maintain U.S. leadership in semiconductors and computers through a 25 to 50 percent increase in pure research, and to add significantly to the supply and quality of degreed professionals. Announced in December 1981, SRC's efforts will focus on *concentrated* long-term research. Its board of directors will be business executives from participating companies in the Semiconductor Industry Association sponsored cooperative. In April 1982, Larry Sumney, who headed the DoD's VHSIC program, became its executive director.

Erich Bloch is Chairman of the Semiconductor Research Cooperative (SRC). He is also Vice President for Technical Personnel Development at IBM. Through his own efforts and those of B. O. Evans, also an IBM vice president, he is leading the company beyond its own corporate confines to join forces with others in the high tech industry. Only a few years earlier, IBM participation in an industrywide cooperative would have been improbable. Today, it is an idea whose time has come. The main reason: Japanese competition.[2]

Bloch's office, in suburban White Plains, New York, is large and comfortable, tastefully adorned by the artifacts collected from the travels of a senior executive. For his visitors, he underlines his viewpoints with data projected onto an enlarged wall screen. It shows the record of Japanese government investments in electronics technology for the past twenty years. Bloch's assembled numbers silently project a picture that still remains officially unstated by corporate IBM. Japan's threat is being taken seriously, and even IBM can no longer go it alone. "They operate in a freer environment," Bloch explains. "There are no antitrust concerns in Japan, the cost of capital is cheaper, and they have learned well the art of cooperative undertakings among companies."

While Bloch appears soft spoken in his presentation of the SRC case and IBM's role in it – "IBM is interested in seeing a viable semiconductor industry survive in the United States in order to avoid the evident problem of relying solely on foreign suppliers that are also competitors," – he is adamant about the means required to fight back. "Both we and the university world are going to have to come together somewhere," he states emphatically. Industry is going to move and others are going to have to move with them. Does he foresee any difficulties in the partnerships between industry and universities? "Surprisingly, the greatest resistance in universities is not from academics but from some of the administrators. If they want support from business they are just going to have to bend. . . . We expect that the need for research support and the need for up-to-date, state-of-the-art equipment may encourage universities toward a modicum of cooperation."

The amount of money foreseen by the Semiconductor Research Cooperative should give food for thought to the planners at the Japanese Ministry for International Trade and Industry (MITI). Twenty American corporations are committed to a long-term research effort that will invest $5 to $7 million in 1982, $10 to $15 million in 1983, and up to an estimated $37 to $55 million per year thereafter to support pure research. Total U.S. research and development in these fields is presently estimated at $600 million, of which only 10 to 15 percent is considered pure research (for example, on materials research or on artificial intelligence). Thus the projected pure research budgets of SRC represent increases initially on the order of 20 to 25 percent and later going as high as 50 to 90 percent. According to SRC's best estimates, these research amounts are almost three times the annual rate in Japan.[3] Since funds will be raised by company contributions proportional to sales, this also means that as industry sales grow, budgetary allocations will also increase in the coming years.

Reminiscent of Japanese strategy, the concentration of research is key to SRC's statement of purpose. "The physical location and the research commitment of SRC should be *concentrated* in major generic areas and institutions rather than spread out among a large number of universities and heterogenous subject areas." The decisionmaking will likewise be focused. SRC's policymaking board will be composed primarily of industry members: one-third elected from the board of the Semiconductor Industry Association with others coming from participating industries and academia. Going after federal funds will be deliberately ruled out, unless they are available in matching grants or government-initiated contracts.

Commenting on the state of affairs of the university world, Bloch points to a long-standing tradition of "independent and competitive research" between universities, which is now threatened by cutbacks in federal government research support. Much of SRC's efforts will in fact substitute industry dollars for dollars the government used to, or should have, put in. A well-directed industrial effort could have significant impact on university research as well as on the vitality of the industry. The U.S. semiconductor industry is at a crossroads today, and sitting squarely at the crossing is the university.

THE MICROELECTRONICS AND INFORMATION SYSTEMS CENTER (MEIS) (University of Minnesota)

MEIS is a located at the University of Minnesota and supported by a consortium of Minnesota-based high technology companies — Control Data, Honeywell, Sperry Rand, Medtronics, and 3M. Founded in 1979, it was among the first university/industry partnerships designed to upgrade EE/CS education and research. It has a special leveraging feature: its $6 million budget must be used for matching grants from other sources, especially government. It seems to be working. An initial university course to receive MEIS funds skyrocketed in enrollments from 25 to 125 EE/CS students in one year.

Halfway across the nation, in Minneapolis, Minnesota, is a man who in appearances is the antithesis of IBM's Eric Bloch. A professor of chemistry and a lifelong academic, Dr. Robert M. Hexter represents the university side of the academic/industrial partnership.

There is no carpet in his office and the walls are too crammed with books and papers to permit screens for projecting data. Professor Hexter can be found in one of the older buildings on the University of Minnesota campus, where the corridors and stairs still carry an atmosphere of academia that harks back to the turn of the century. This is a no frills academic department office. A few too many boxes of documents overflow the cramped shelves; the secretary seems to be handling several jobs at once, so that when the phone rings, it is not clear who should attempt to answer. This is the world of budget cuts, endless proposals for grants, and departmental academic infighting. The calm pace of an industry office contrasted with the hectic daily life of a university department forms an all-too-telling contrast. Shouldn't the image be the other way around? It is not surprising, the visitors reflect silently, that so many academics can be lured away to more lavish laboratories and higher salaries in industry.

But Professor Hexter is a dedicated man. Driven by concerns that transcend his own professional field of chemistry, he represents in many ways an emerging academic style that is more enterprising than the conventional expectation held by many industry leaders like Bloch. Hexter is a leader from academia who has actively sought a partnership with industry. Combining forces with Walter Bruning, vice president for data services at Control Data Corporation and formerly a vice president of the University of Minnesota, he was able to get the support of Control Data's chairman, William Norris. In late 1979, a new venture was started with a $2 million challenge grant from Control Data. "We quickly found the other $3 million from industry," said Hexter, "and we were up and running by the end of 1980."

The Microelectronics and Information Systems Center (MEIS) is not lavishly endowed and cannot indulge in the apparent security that surrounds the Semiconductor Research Cooperative. But one fundamental similarity — concentration and focus — had to be learned the hard way. In Dr. Hexter's words, "our first idea was to be a mini National Science Foundation soliciting proposals. But before you knew it, we had as many ideas as committee members. There were twenty-one members and twenty-one ideas, and some were even proposing support for their own work."

As a result, a loosely structured committee and broadly defined program were quickly turned into a tightly managed venture. Regrouping the MEIS under a management team of nine members, of whom four were from industry and five from the university thereby giving an important but essentially symbolic control to the university side, the focus of the MEIS was narrowed to four key programs.

"Instead of being in twenty-one fields, we chose those in which we felt the university was genuinely strong or was at near full strength," recounts Hexter. These include the expansion of a microelectronics laboratory and construction of a design automation lab, the strengthening of a software engineering program, and two staff programs to increase faculty size and to create an MEIS Fellows Program with a high prestige profile "to help in catalyzing a key group of individuals in the EE/CS field."

As an extra twist, the MEIS is applying the fine art of leveraging its privately supplied initial funding source. None of its initial capital of $6 million will be used for building, and what moneys are available can only be applied toward matching grants. For example, there is a proposal to the Army Research Office in Durham, North Carolina, for $750,000 for each of three years to be matched by MEIS. In effect, Hexter's experiment is to turn one industry dollar into two or more; and the outcome is designed not only to enhance MEIS as a research center but to increase significantly its teaching and enrollment capacity.

In another way Professor Hexter would surprise Erich Bloch's cautious view of academia. "We are not so interested in hands-off grants from industry," explains Hexter. "We want to work directly with a company so that there is an active two-way exchange." This is visible in the already expanded computer-aided design program that increased its enrollment from 25 to 125 students as a result of an equipment grant from the Calma Company, a division of General Electric. Many of these students spend part of their hands-on laboratory hours on site at the corporation. While a logistical disadvantage, the interchange of academic people into industry and of industry people into the university is a healthy phenomenon, argues Hexter.

THE CENTER FOR INTEGRATED SYSTEMS (CIS)
Stanford University

Founded in 1981, the CIS will be the first university-based laboratory with a capacity to manufacture Very Large Scale Integrated (VLSI) circuits. The CIS is funded by a $13 million fund pooled by a seventeen company consortium.* An additional $8 million will come from a contract with the Defense Advanced Research Projects Agency (DARPA) of the Defense Department. The Center will combine electrical engineering and computer science fields and will turn out 100 masters and 30 doctorates a year. Thirty faculty members from various departments will be affiliated with the center. Its $14 million laboratory facilities will open in 1983.

* General Electric, TRW, Hewlett Packard, Northrop, Xerox, Texas Instruments, Fairchild Semiconductor, Honeywell, IBM, Tektronics, DEC, Intel, ITT, GTE, Motorola, Monsanto, and United Technologies

John A. Young is the chief executive of Silicon Valley and Stanford University's prize child – the Hewlett Packard Corporation. His trappings and management style are disarmingly informal, acquired from a corporate culture that thrives on maintaining a nonhierarchical work environment. The single story building[4] that houses Young's office is open, unbroken by anything more than plants and shoulder-high partitions for its entire length, letting in a flood of pleasant natural light. That the walls to his office are simple partitions is as striking a contrast to visitors from the East Coast as is the physical contrast between sunny, warm Palo Alto and Minneapolis-St. Paul, where office complexes are interconnected by "winter-proof" glass walkways.

John Young has just raised $13 million for the Center for Integrated Systems (CIS) at Stanford University and he is clearly pleased. His success has a twist that he describes with pride. "We managed to get IBM to do something they seldom do, . . . help pay for bricks and mortar for a new research facility."

A new building, to be located on the Stanford campus, will house the center. Research will focus principally on Very Large Scale Integrated (VLSI) technologies. It will be administered as a semi-independent center, a co-venture between industry and the university. Seventeen companies each contributed $750,000 to be paid over a three-year period. These funds will pay for construction and equipment; and $4 million from a DoD contract will help defray building and equipment costs.

Dean William Kays, an active partner in CIS, sees the program as pushing Stanford engineering "into the world forefront." Part of this ambition includes creating a "new kind of student with a combined competency in computer sciences and electrical engineering." This goal is shared by industry sponsors, who see their payoff not merely in terms of new research but especially in access to a new generation of scientists. "What they can get," says the center codirector James Meindl, "can be encapsulated best by the words 'lead time'. They get a good window on more than $12 million a year research program."

For Meindl and codirector John Linvill, the new center represents a vertical integration of university research. "You go from a material to a device to components to systems. Just gearing up a university research activity to achieve that kind of integration takes some dialogue and a tremendous amount of it." To one of the first corporate participants, George Pake of the Xerox Corporation's Palo Alto research laboratories, the breakthrough in getting various parts of a university to work in concert was the real achievement. "Perhaps we should have called it the Integrated Center for Integrated Systems instead," he jokes.

While the CIS is not unique in its genesis as a corporate and university partnership, it stresses more than other ventures a central role for the federal government. CIS looks to Washington for a major source of operating funds. "You must remember," says Kays, "that Stanford depends on the federal government for 99 percent of its research funds. We expect the new center to be sustained by contracts from Washington."

Stanford's Center for Integrated Systems brings together industry, university, and government in VLSI research collaboration. The industry initiative, personified by John Young's aggressive fundraising for the CIS, is directly linked to Stanford through a hybrid research facility. It reports directly to the university provost and belongs to no single academic department. Its research is informally monitored by an advisory council made up of faculty and sponsor-company representatives; all of the latter are provided with office space in the laboratory. This allows industry to play a more influential role in the life of the research center than ever before. And with government funds playing a major role in sustaining the operations, the net result is a de facto three-way partnership. One might call it a plan without national planning, a scheme that needed no Japanese MITI or French centralized agencies.

THE ELECTRONICS EDUCATION FOUNDATION (American Electronics Association)

The Electronics Education Foundation is a nonprofit organization set up and controlled by the California- based American Electronics Association (AEA). Its purpose is to pool money contributed by member companies of AEA to improve engineering education. Contributions are targeted at 2 percent of R&D, and AEA companies represent sales of about $77 billion and R&D of nearly $5 billion. A fundraising target is set that will allow the AEA to disburse money widely in the form of 300 grants to raise faculty salaries ($10,000 each), 200 graduate student fellowships ($15,000 each), and a variety of other "pass through" grants to U.S. colleges and universities. Its first operating year was 1982 with a start-up budget of $50,000.

Dr. William Perry, former Chairman of the AEA's Board of Directors and prime mover in the launching of the AEA's Electronics Education Foundation (of which he remains chairman) was formerly Assistant Secretary of Research at the Defense Department during the Carter administration. His private office at the staid investment banking firm of Hambrecht and Quist in San Francisco is decorated with military memorabilia — medals, citations, war plane pictures, and political handshake photos. His concerns, though softly spoken, nonetheless convey a tone of urgency: "Our projections of engineering shortages only scratch the surface," he says. Referring to the AEA study that graphs the shortages in EE/CS degrees: "These are the most conservative numbers you could come up with. The problem is far more severe than our numbers suggest. We don't even address the needs of other industries."

William Perry was an architect of the DoD's Very High Speed Integrated Circuits (VHSIC) program and is now an advocate leader for industry. More than just a lobbyist for its interests, he is an activist catalyzing its members not only to act, but to act quickly. The week of our visit, William Perry and Patricia Hubbard, the AEA vice president for engineering education, were winning the battle for approval to establish a foundation. Established in September 1981, it represents a frontal attack on the national dilemma of engineering departments — lack of money.

"We've set as a goal contributions by high technology industries equal to 2 percent of their current R&D budgets," said Perry. "Our membership has more than $77 billion in total sales and about $5 billion in R&D. The industry as a whole is three times that size. We see a potential amount from our members alone that could reach $100 million dollars. . . ." The foundation's major short-term aims are to increase the number of faculty in engineering departments; to increase graduate student supply; to increase and upgrade equipment and facilities; and to undertake demonstration projects using media-delivered instruction such as ITV.

Foundation president Pat Hubbard stresses that they operate on a regional basis with committees initially in Arizona, Colorado, several California counties, Oregon, and Washington with liaison to the Massachusetts High Technology Council located near Route 128. Another important facet to the AEA Foundation is its focus on universities that produce electrical engineering and computer science graduates but which may not be located near major employers. "Only four of the top ten BS/EE- and BS/CS-producing states have concentrations of electronics companies," says Hubbard. "The foundation will have to help in directing funds from companies with a national perspective to engineering colleges in states such as Pennsylvania, Illinois, Indiana, Michigan, and Ohio."

The AEA Foundation reflects the traditional voluntarist spirit in America, which again is being pressed into action in reaction to governmental budget cuts. Also, following a long-established American tradition, and in contrast to the Semiconductor Research Cooperative, the Foundation will spread its funds over many institutions across the country. It thus fulfills a basic mission, offering the possibility of much-needed support on a very broad front. How successful it will ultimately be will depend on the ability of its leaders to marshal industry support and channel it to the schools that can use it best.

THE WANG INSTITUTE OF GRADUATE STUDIES
School of Information Technology

The Wang Institute of Graduate Studies is a leading-edge experiment. Neither a trade school nor a feeder for Wang Laboratories, the institute attempts to surpass Harvard and MIT in providing master's level education for a high tech economy. Graduating its first five students in 1982 with expectations of 200 in the next five years, the institute provides an education in both management and technology. Faculty salaries are set high, at industry equivalents. Equipment is new and computing access virtually unlimited. The curriculum is international, with European guest lecturers and Japanese management material. Dr. An Wang is the institute's founder. His contributions, plus those of his family and some fifty companies and other individuals, support the institute. The initial capital investment was $4 million and the annual operating costs are currently $1.2 million.

The setting is picture book New England. Tyngsboro, Massachusetts – a curving river running through a gently sloped wooded countryside, the graceful reflection of the arching Tyngsboro Bridge, a white spired village church. Off a narrow road is a 200 acre estate, once tended by members of a Marist order. Once inside the main building, a surprise awaits the visitor. Instead of religious figures and an ornate altar, the one-time chapel is now adorned as a modern high technology center of learning. Computer journals line the walls. Students sit intently reading the latest in software developments. Below them, in a dust-free glass-walled room, are three computers purring nonstop. A Digital VAX 11/780, a PRIME 750, and a WANG VS. These, plus two other WANG systems, support a network of various terminals and printers throughout the one-time seminary, rechristened in 1979 the "Wang Institute of Graduate Studies."

What is unusual about the institute is not just its quaint juxtaposition of high technology and religion. What is special is that it represents education reaching into new frontiers. In 1979, Dr. Wang, believing that traditional academia would be adverse to founding a new professional degree granting program, decided to go it alone. "In universities," said Dr. Wang,[5] "salaries are fixed and that means no competition. And with tenure, professors don't always have incentives to do their best work; professors, if they are good, don't care about tenure. Plus," he added, "we wanted to get started with our program very fast. University bureaucracies don't react fast." Emphasizing the last point, Dr. Wang reminded his visitors that what was just an idea in February 1979 became a working academic institution only eleven months later. What traditional academia could not do for the training of a new generation of computer specialists, Dr. Wang took the initiative to do himself.

The charter class of five students graduated in August 1982 with master's degrees in software engineering (M.S.E.). Their faculty consisted of three full-time members and a number of visiting lecturers. The current enrollment of about thirty students has unlimited access to the latest computing equipment. This puts the Wang Institute into a coveted national position as a state-of-the-art educational center.

Several characteristics make the institute an experiment of significance. One is its emphasis on high-quality teaching at the leading edge. William M. McKeeman, chair of the faculty, says:

> One reason we are not at Harvard is that their faculty is pressured to publish or perish. Here we concentrate on developing and teaching new material because the field of 'software engineering' is new. And we are convinced that by offering salaries equivalent to the best in industry, we can attract the top-rated faculty necessary to make the Institute work. Harvard would not allow that.[6]

McKeeman, recently moved from the University of California at Santa Cruz and Xerox Palo Alto Research Center, exudes genuine enthusiasm for his new job and his new academic responsibilities. He sits in a modern, well-furnished but small and unostentatious office. Behind him lie various lengths of cable that hook up two separate terminals to the hardware humming away in the basement.

Aside from high salaries, another aspect of the institute's program is particularly important: developing a discipline of software engineering that includes management and technology. Says McKeeman:

> We are trying to form a discipline by working 'in the trenches.' Software engineering concerns itself with the very simple question of how you get the science of computing done. We worry about the human factor — especially how to train software specialists to function in groups. We also worry about how to build a software system that is readily fixable. Think for a moment of software as if it were an engine. We want all the bolts and fan belts to be easily accessible for quick and simple repair.

To make his point, Professor McKeeman reminds his visitors that in the world of computing, the main cost of software is not in its running but in its failure to run smoothly. The unanticipated expenses resulting from crashes and lost time are the greatest danger. "Our job is to make the use of software equivalent to a flawless train ride rather than the experience of a train crash."

Students at the institute are eligible to apply only if they have a minimum of two years of software employment experience and a professional working knowledge of computer languages and software. "In fact," says Cynthia Johnson, the institute's corporate liaison, "of our current enrollment of five full-time and about twenty-five part-time students, work experience averages 4.9 years and the average age at the time of application is twenty-nine. Because of their age many of the students do not have many family commitments and can devote long hours to their school work."

Most of the current students are sponsored by some fifteen local companies where they are employed. On-campus recruiting of company sponsored students is prohibited — otherwise the school might be seen as a way to switch employers. The five-year goal is to build a student body to a steady level of about 200. "We want to create a new professional school," says Professor William McKeeman. And under his direction the institute is covering new ground. Only Seattle University in Washington State has a similar attempt under way since 1979 with the support of Boeing Aircraft Corporation.

The pioneering efforts of Dr. Wang mirror an entrepreneurial flair that is not unlike that of the great fortune builders of the previous century. As Walter Saxe, the financial vice president of the institute puts it, "the institute is a result of philanthropy that is reminiscent of a Carnegie — but in a different time." Dr. Wang and his family are the main financial supporters of the institute, which runs at an annual cost that in 1982 was $1.2 million. Because the start-up enrollment is low, tuition makes only a small contribution to this figure. The initial investment to found the Wang Institute totaled about $4 million. About $600,000 bought the estate. An educational bond issue raised $2 million for physical renovations. About $1 million in equipment was acquired at a cost of $200,000 — the balance being contributed by major manufacturers.

Reflecting on the institute ten or more years from now, Dr. An Wang sees a broadening of its academic horizon. "Well, we hope to have more programs. The School of Information Technology is just a first step. But we do not want to be limited to technology." Listening to Dr. An Wang, one cannot help thinking how quickly he, his company, and his ambitions have come of age. As Robert Saulnier, one of his first employees and now an aide, recalls: "Only fifteen years ago we were selling calculators to schools and weren't quite sure we'd make it." Sales have now reached the billion dollar mark. The commitment to found an academic institution punctuates the maturing not only of a company and its founder, but of an industry.

Leaving the Wang Institute, one winds back along the riverside. Not far up the road is a crowded and unkept, almost ramshackle, truck-stop diner. To enter is to move back to another time when the Merrimack Valley was the world center of textile production and machine tool shops. The patrons in the luncheonette are all men who labor with their hands. They wear grease stained overalls; their hands and faces are burnished by years of hard work and by the oily dirt of machines. The contrast between the new knowledge-intensive software researchers at the institute with its dustless humming computer rooms and the workers symbolizing an older industrial age is a fitting reminder of a historic transition. The Merrimack Valley has entered a new age.

THE MICROELECTRONICS CENTER OF NORTH CAROLINA (MCNC) (Research Triangle Park)

MCNC was incorporated in July 1980. Located in the Research Triangle Park midpoint between Raleigh, Durham, and Chapel Hill, North Carolina, it was launched in 1981 with a grant from the state legislature for $24.4 million to cover construction, equipment, and initial operating costs. Its purpose is to achieve a quantum jump in training and education, at all levels, in order to prepare North Carolinians for employment in the nation's fastest growing industry, and to make North Carolina competitively more attractive for employers than are other states and regions. MCNC will work closely with five North Carolina universities and the Research Triangle Institute, as well as with industry — especially the nearby $50 million microelectronics production facility just set up by General Electric.

Governor James B. Hunt is an aggressive and successful promoter of North Carolina as a home for a new generation of high technology industries. He personifies the public sector at its best in conceptualizing and moving toward a future-minded growth strategy. James Hunt is proud of his efforts to help launch a microelectronics research center of major national significance, and with an appropriation from the state legislature of $24 million in start-up funds. As his administrative assistant, Barbara Buchanan warned, "you'll end up talking with him far longer than you think if you get him on that topic."

The Governor's Office in the small, historically quaint granite State House, in many ways reflects Governor Hunt's personal style. He seems unfettered either by the state's operating bureaucracies or by the legislature, physically separated into another building across a wide avenue. In the small, almost intimate spaces of the renovated State House, Hunt seems at home. And he is very direct about how he perceives his responsibilities. "Only the governor can bring it all together. I doubt you'll ever get maximum effort without a strong governor. You need the leader who will go out and take the initiative to get more resources."[7]

What quickly becomes apparent in a conversation with James Hunt is not only a demonstrated ability to lead but also a rare political quality of having a lucid, over-the-horizon view of what needs to be done. "North Carolina needed a strategy for economic growth. And we are now doing this by concentrating on areas of good potential for the future. Our commitment to education and growth makes us a visionary state." The state, which is forty-first nationally in per capita income standing, is ninth in per capita expenditures on higher education. This commitment is the heart of the governor's strategy.

More than twenty years of slow and steady efforts to modernize the economy have come to fruition in the Research Triangle Park. From modest beginnings in the late 1950s, the state labored steadily in forging an alliance between industry, academia, and government. The success of these efforts is catching other high technology regions off guard in the national competition for new growth and more jobs.

Much of Hunt's success is due to the close working relationships he has developed. Relying on a Science and Technology advisory board founded by former governor Terry Sanford as a consensus-building entity, Hunt has followed their recommendations in identifying areas for economic growth such as microelectronics and biotechnology. In addition, the governor has maintained a "day-to-day relationship" with the state's university presidents. "I view the universities as service organizations to help us grow economically." Hunt maintains the same close relationship with the business community. Thus when General Electric came to North Carolina in search of a possible site for an electronics research base, the governor was ready. He was able to offer a state in which government and educators were primed to enter into partnership with business.

"We saw a strong future only if the 'people' would be ready to provide the facilities to encourage active cooperation between government, universities and industry," said Hunt. This was demonstrated, convincingly in his view, by the legislative appropriations that culminated two decades of effort by North Carolina to capitalize new growth opportunities. These were fortuitously timed to coincide with an initial $55 million research center commitment by GE — a commitment the company doubled in 1982. These commitments raise to more than $1 billion the value of new buildings in Research Triangle Park alone — the core of the diminutive but Ph.D.-rich tri-town area of Raleigh, Durham, and Chapel Hill.

Such investments rapidly moved a unique conception of an industrial park into national prominence. Research Triangle Park, a 5,500 acre nonprofit undertaking, is the center of gravity of a strikingly small metropolitan area. Within a ten-mile radius of the park are no more than 400,000 permanent residents. Yet 20,000 people work at the Research Triangle with companies that are a growing directory of star performers: IBM, Northerm Telecom, Burroughs Wellcome Company (pharmaceuticals), Data General, and Monsanto Triangle Park Development Center, Inc. The locally developed Research Triangle Institute founded in 1959 now counts 1,000 employees.

To a first-time visitor, the scale and visual impact are unexpected. With only 15 percent of the land-surface in the park open to development, the first impression is of wide expanses of forest interrupted here and there by stylishly modern complexes of research and manufacturing companies. It has none of the chaotic and feverish tempo of Silicon Valley, or the more austere and businesslike aura around Boston's Route 128. Yet with a network of three universities linked to the Park's activities and an atmosphere of commitment to state-of-the-art research evidenced by the legislature's funding of MCNC, one cannot but feel the gravitational pull. In the words of George Herbert, president of the Research Triangle Institute and board chairman of the microelectronics center:

> The MCNC was not created as a subsidy to industry. Rather it is intended to support the educational and research missions of the six participating institutions[8] to prepare greater numbers of people for careers in high technology industry. This creates a magnet effect to assure that opportunities for those careers are created in new industry here in North Carolina.

Herbert notes that the starting budget will be about $10.5 million for the MCNC facility, including preparation for space at the campuses, $8.6 million for equipment, $2.8 million in direct operating costs, and another $2.5 million for programmatic expenditures and graduate fellowships at participating academic institutions. In addition to the state funds, MCNC intends to raise more financing from nonstate sources, especially industry. The educational impact, he estimates, is that graduate enrollments in electrical engineering and computer sciences will jump nearly 50 percent, from 273 to 400, in the first year alone. The economic impact is measured in employment. GE officials see initial employment in the 150 to 200 range, and jobs could grow to 1,000 if operations go well.

NORTH CAROLINA

A NEW HIGH SCHOOL GOES HIGH TECH

The North Carolina School of Science and Mathematics is the nation's first residential public school. It was created in 1978 by the North Carolina Assembly at the suggestion of Governor James B. Hunt, Jr. It is a statewide, coeducational school developed specifically to provide a challenging education for eleventh and twelfth graders who are gifted in science and mathematics. The school is an important step toward reversing devastating national trends — the neglect of education of high caliber students, the decline in quality of science and mathematics education, and reduced economic productivity.

The school opened with its first class of 150 eleventh graders in September of 1980. Enrollment is planned to increase by 150 each year to a maximum of 900 eleventh and twelfth graders. The student population is 50 percent male, 50 percent female; 24 percent are from minority groups. Its future success will be a direct result of an extraordinary partnership between the public and private sectors.

The school has attracted strong outside support from foundations and corporations. Computers are an integral part of the science and humanities curriculum of the school. "Each graduate of the two-year program will have used the computer as a tool," say school founders, "they will have an appreciation of the uses and limitations of computers, and will have written several programs." An attractive inventory of equipment — much of it donated by industry — makes the task feasible: five Apple IIs, one TRS-80, and four terminals hooked into the area's university facilities. A DEC VAX-11/750 is being readied for installation by next year.

Source: Paraphrased and quoted from Steve Davis and Phyllis S. Frothingham, "Computing at a New Public School for Gifted Students," *On Computing Inc*, Spring 1981, pp. 72-75.

The efforts in North Carolina are not limited to higher education. The state has also started a new residential high school in Durham. Reminiscent of the famed Bronx School of Science, the North Carolina School of Science and Mathematics caters to gifted students in math and science. Only two years old and with an enrollment of 300, the school is a pioneering venture as the nation's first public residential high school. It is the first of its kind in the country and demonstrates the commitment by the state and the governor to build a human resource strategy as a foundation to the region's new economic growth.

Building Momentum for Change

The six cases just outlined, while individually prominent and far-reaching, are but a few examples of building a momentum for change. As a multitude of projects, legislative measures, and new educational ventures are launched, a sense of regional pride and competitiveness grows. Not to be outdone by competing regions, different states in the country are proposing further initiatives. Massachusetts is a case in point.

Business leaders and leading academics in Massachusetts still smart at the realization of their failure to capitalize on silicon technologies a generation ago. Instead they watched the rapid growth of these technologies blossom in California. But by 1982, the lesson seems to have been learned. George S. Kariotis, Massachusetts Secretary of Economic Affairs and an active spokesman for the electronics industry, states the matter bluntly. "Massachusetts is in the minor leagues in semiconductors. The West Coast out-hustled and out-maneuvered us. Now we've got to do what we can to hang on to our semiconductor companies and develop them further."[9] What triggered Massachusetts into action, says Kariotis, was the sudden realization that states like North Carolina were actively funding projects like the microelectronics center with major amounts of money.

Spearheaded in large part by industries in the Massachusetts High Technology Council (MHTC) and by a subsequent marriage of interest among eight leading universities and colleges,[10] Kariotis has been able to forge a state legislative program. The Massachusetts Technology Park Corporation, a $40 million semiconductor research and education laboratory with an estimated $3.5 million annual operating budget, has been proposed, and legislation for a $20 million bonding authority has been agreed to by otherwise feuding political leaders on the condition that a matching amount be funded by industry.

Looking over his twenty-first floor vista of Beacon Hill and the Charles River, Kariotis reflected on the difficulties of pulling all the pieces of the partnership together. "The key," he says, "is that we achieved the unbelievable. We got all the parties talking constructively to one another for the first time."[11]

The state initiative in Massachusetts is paralleled by an equally important effort by MIT to maintain its own position in semiconductor research. The institute is only $5 to $7 million dollars short of meeting its $21.1 million goal to fund a new VLSI research center, formally part of a larger entity, the MIT Microsystems Program. Similar in some respects to Stanford's Center for Integrated Systems, the new MIT program is an institute-wide one, administered by the Electrical Engineering and Computer Science Department. Its director, Professor Paul Penfield, Jr., also coordinates a Microsystems Industrial Group of corporate participants. Corporate sponsors will be members of an advisory panel and will have access to professorial colleagues and to student research.

To Professor Richard B. Adler in MIT's Electrical Engineering and Computer Science Departmemt, the new program is urgently needed to stay on the frontier of work on the circuits and systems of today's electrical engineering. Commenting on the difficulty of foreseeing the direction of microelectronics research over the next twenty years, he noted that "whatever happens, we had better be in this now. If we miss the boat, leap-frogging will be impossible."[12]

This view is echoed by the head of the department, Professor Joel Moses. "How should we deal with the fact that many of the ideas for the Fifth Generation Computer Project [now located in Japan] came from MIT? We explained them to American industry but they wouldn't go for it. The importance of parallel processing and artificial intelligence was not fully recognized by IBM and the other American computer companies. Instead, Japan was quicker to see the application possibilities. Now that Japan has announced its intentions, however, we begin to see some movement by American industry in these areas."[13]

Does it all add up?

What do we learn from examples such as those just described? On one hand they reassure us of the vast pool of entrepreneurial vitality that exists in American institutions. They also point to the rich potential for collaborative experiments among industry, university, and government in ways that are only just beginning to bear fruit.

The opportunity is succinctly framed by Dr. An Wang. "In order to prosper, a growing partnership for industry and government is important. Every partner must not only get something but must return something. One cannot look at the other side as an adversary."[14] One of Dr. Wang's aides, Paul Guzzi, illustrates the point by saying: "Wang Labs built and expanded by direct actions that in turn led government to be as encouraging and supportive as possible. An unspoken trust developed. In Lowell this trust includes the people of the city whose economic self-confidence is involved."[15] Such give-and-take partnership between industry and government may not be novel, but it is far from common.

On the other hand, does the total of individually initiated state, corporate, and academic efforts add up to a national response adequate to the challenges of the 1980s? The answer is probably no.

The new partnerships probably raise as many questions as they answer – about who can and should initiate and guide such projects, who can and should pay for them, and who can and should provide some sort of national coordination and leadership for new educational and economic responses. They also bring into focus the incredible contrast between the cash-rich high tech industry and an underfunded system of higher education struggling to make ends meet and to keep afloat.

In the next chapter, we investigate some of these issues – the reasons behind the declining resources for education, possible new roles for state government and for industry, and some of academia's worries over potential problems raised by corporate financing. We will also look at some new initiatives that the academic world can undertake – most notably, moves toward lifelong learning and the use of instructional television – to make higher education more responsive to new needs and also to ensure its own survival. The educational world is in trouble, and it knows it.

Chapter 7

THE STRUGGLE TO STAY AFLOAT

STORMY IN the 1960s, subdued in the 1970s, American education is now in disarray. Inflation, declining demographics, withdrawal of federal support, and a public mood of disenchantment have combined to work against the schools, whether public or private, high school or college. All of the trends and numbers point downward, except one – the cost. Presidents and chancellors of the nation's nearly 3,200 institutions of higher education are learning to share an increasingly common concern, namely, how to stay afloat. More difficult still, a predicament of double dimensions has emerged: how to stay afloat and still respond to the changing demands of a new economy and knowledge-intensive society.

In Massachusetts, MIT President Paul Gray says that the system has reached a "saturation point." Worcester Polytechnic Institute (WPI) President Edmund Cranch foresees a "national crisis" in education. And Northeastern University's President Kenneth Ryder says his two colleagues "underemphasize the dimensions of the problem."[1] Let's look at some of the dimensions of decline.

Where More Means Less

While the resources that support higher education are rising in inflated dollars, the percentage of GNP devoted to education has declined from a high of 8 percent in 1975 to around 6.9 percent in 1980.[2] In 1960, the bill for American higher education was $6 billion; in 1970 it was $21 billion; by 1975 it reached $39 billion, and by 1980-81 higher education was a $65 billion enterprise. Thus, while the dollars seem to rise, the net effect is decline. This would seem to be the wrong direction for a society headed toward an era of greater knowledge intensity.

In the 1980-81 academic year, public institutions accounted for $ 43.5 billion, or 67 percent, of the total expenditures, while private institutions accounted for $ 21.5 billion or 33 percent. For public institutions, the largest share of revenue comes from state and local governments (50.9%); for private colleges and universities, most comes from tuition (37%). Support from the federal government is roughly the same for each sector – about $4 billion to public and about $3 billion to private institutions.[3] It is these amounts that the Reagan Administration plans to trim. While federal support for higher education represents only fifteen percent of the total, it is critical for two reasons – they are targeted to research and development and there are few alternative sources that could supercede federal funds at present levels.

DOLLARS FOR SCHOLARS:
Sources of Funds for Higher Education

"Current Fund Revenue"
Figures for 1978-79 (in billion of dollars)

	public	%	*private*	%
Tuition & Student Fees	$ 4.1	13.1%	$ 5.7	36.9%
Federal Government[a]	4.0	12.8	2.9	18.9
State Government	14.4	45.7	0.3	2.0
Local Government	1.6	5.2	0.1	0.7
Private Gifts	0.8	2.5	1.6	10.0
Endowment	0.1	0.4	0.7	4.5
Sales & Services[b]	6.3	20.2	4.1	26.9
TOTAL	31.5	100.%	15.5	100.%

Notes: [4] [a] For both public and private, the majority of the federal money goes for "restricted grants and contracts."

[b] "Sales and Services" is primarily auxiliary enterprises such as room, board, and activity fees including sports, but it also includes income from university hospitals.

Source: *Digest of Educational Statistics*, National Center for Educational Statistics, Washington, D.C., 1981.

Federal Cutbacks

While justified in some cases, Reagan's efforts to strip away fat may end up bleaching higher education to the bones. Appropriations showed an 11 percent drop in financial aid and a 30 percent drop in student loan subsidies between 1981 and 1982. But even more important is research support and the need to increase rather than decrease federal funding of university research.

The R&D activities of leading research universities depend heavily on federal funding. About 80 percent of university R&D is carried out in 100 so-called "leading research universities."[5] These universities grant more than half of all graduate degrees in science and engineering and are thus

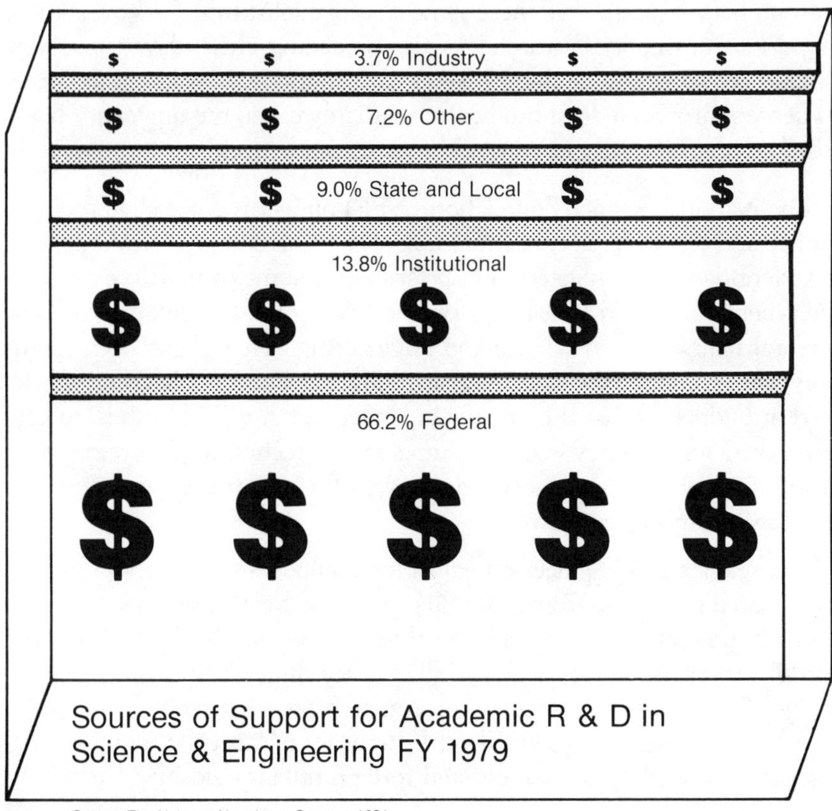

Sources of Support for Academic R & D in Science & Engineering FY 1979

Source: The Nchems Newsletter, Summer 1981

critical to the vitality of the high tech industry. The top universities derive almost 70 percent of their R&D funds from federal sources. Between one-third and one-fifth of their total edcuational and general revenues also come from the federal government. Of the $5 billion university R&D expenditures in science and engineering for all universities in 1979, 66 percent came from Washington. State and local government shouldered only 9 percent or less than $500 million of the university R&D bill. Less than $200 million, or 3.7 percent, came from industry sources.[6]

It is healthy in many ways for industry to get more involved with universities and to contribute substantially more support to university research. Tax credits for industry support of research and equipment grants in the 1981 Tax Act are an important recognition of this fact. But the numbers indicate that there is no feasible substitute for government support of university research. As the preceding chart shows, industry would have to nearly double its R&D contributions to make up for just a 10 percent drop in federal funds. Rather than weaken the university R&D resource, the national goal should be to strengthen it.

The National Science Foundation (NSF) budget is especially crucial to science and engineering education. Under the Carter administration, NSF had identified and budgeted high priority programs to update outmoded equipment in university laboratories and to establish a number of new doctoral fellowships in science and engineering. These have been essentially eliminated in the new Reagan budgets. The double irony of the current budgets is that the greatly expanded defense R&D funds, mostly earmarked for industry, will consume scarce technical resources, while shrinking NSF funding, mostly earmarked for universities, will generate less new technical manpower.

The science and engineering education component of the NSF budget is estimated to decline from $70 million in fiscal year 1981 to $20 million the next, down to $15 million in fiscal year 1983. By the fall of 1982, the overall decreases will exceed $1 billion according to the Congressional Budget Office, a total that is underestimated by 50 percent in the judgment of the American Council on Education. Meanwhile, defense procurement jumped by 35 percent and foreign military aid by 51 percent.[7]

The science and engineering education component of the NSF budget, falling to about $15 million in 1983, is almost seven times less than the

$111.9 million planned by the Carter Administration. Some project areas will see their funding totally eliminated.

DOWN TO ZERO: Selected NSF Cuts

(figures in millions of U.S. dollars)	1981	1983 (est)
Scientific Personnel Improvement	$33.4	15.0
Science Education Resources Improvement	17.8	0
Science Education Development & Research	10.9	0
Science Education Communication	8.6	0
TOTAL Science and Engineering Education	$70.7	$15.0

Source: National Science Foundation.

Other budgetary items that directly affect science and engineering educators are being subjected to drastic, and in many cases total, cutbacks. Thus $6 million in funds for research libraries in 1981 is slated for elimination in 1983. College library funding of about $3 million would also disappear. However, in such fields as agriculture – with a strong lobby – the library budget will actually increase from $8.75 million to over $9 million by 1983. Such funding inconsistencies are one reason why some legislators such as Congressman Edward Markey from Massachusetts see them as analogous to "book burnings."

What does Reaganomics mean to a university president? At Northeastern University, one of the nation's largest private institutions with a significant annual output of engineering graduates, President Kenneth Ryder estimates that proposed cuts will cost his students, and ultimately the school, more than $16 million in student loan aid, and another half million in lost state aid. This is serious for a university like Northeastern, which depends on tuition fees to meet 75 to 80 percent of its annual expenses. These numbers are too large to be easily made up from other sources. Boston's high technology companies, for example, would be lucky to raise $14 million a year – and this to be shared among fifteen to twenty schools in the region. As Edward E. David of Exxon Research notes, "the federal budget is slipping to such a point that even if industry tripled its giving, it would not make up for one year's cut in federal government support."[8]

Demographics

No discussion of higher education would be complete without reference to demographics. Demographic shifts reinforce – perhaps overly so – public perceptions of an expected decline in higher education costs. Indeed, it is sobering to contemplate the rapidly decreasing pool of high school graduates available to enter a college or university. Nationwide, the drop is estimated at 26 percent by 1991. In some regions, the decline is even more severe. In New England, long a national leader in education, and in Massachusetts, home of numerous and prestigious educational institutions, the projections are among the worst: by 1995 a drop of 37 percent for New England and 42 percent for Massachusetts.

DEMOGRAPHERS' DREAMS, EDUCATORS' NIGHTMARES

Numbers of High School Graduates & Projected Percentage Declines

Year	1979	1987	1991	1995
U.S.	2844	2376	2131	2279
		-16%	-26%	-22%
New England	164	132	106	103
		-19%	-35%	-37%
Mass.	79	63	49	46
		-20%	-38%	-42%

(Figures in thousands)

Source: "High School Graduates: Projections for the Fifty States," Western Interstate Commission for Higher Education, Boulder, Colorado, 1979.

Such projections serve as a convincing and often used excuse to withdraw resources from education. While this could be justifiable at the elementary and secondary levels, the case is less clear at the post-secondary level. How well do we really understand the impact of the shift toward knowledge intensity on the demand for higher education? Will a significantly larger number of high school graduates go on to college in order to qualify for employment? What will be the impact of lifelong learning requirements, especially for scientific and technical occupations, on the capacity of the higher education system? Capacity requirements could in fact increase with a declining population, although perhaps not with the same mix of fields or age groups that are the norm today.

What is clear is that the number of unskilled jobs that require little or no education will be in continuous decline in America over the next decades. This is especially true in manufacturing industries. Multinational companies seeking to optimize the deployment of resources will take advantage of the ability of Third World countries to compete successfully in labor intensive operations. For Americans, the best alternative is to provide educational support rather than welfare payments. So far, the formulas for such programs have failed. Rather than quit, we should try harder and learn from past mistakes. One of these mistakes was a failure to significantly link the goals of higher education with the goals of economic development.

Two opposing trends – rapid growth of the high technology industry and the expected rapid decline of the high school age population – will have an important impact. Barring radical new productivity changes, a greater percentage of our high school graduates will have to take up engineering studies if employment demands are to be met. This in turn will require more competence in science and mathematics by a greater number of high school students, including women and minorities.

While the pressures toward more engineering are real enough, the magnitude of the shift required is often exagerated in discussions of appropriate goals for higher education. Only 6 percent, or 60,000, of the almost one million bachelor degrees awarded annually are for engineering. Even if the number of engineering degrees were to double, it would be well short of the competition in Japan and West Germany, which respectively produce 21 percent and 37 percent of their degrees in engineering. While the numbers are not so large in terms of total output of higher education, their leverage and cumulative impact on the economy are enormous.

Who Can Help?

Nearly every university president and higher education official is asking the same question – who can help fill the gaps in shrinking resources? A major unknown is what state governments will do about support for higher education under Reagan's "new federalism." The mix between public and private varies considerably by state and by region. For example, in Massachusetts 80 percent of the graduates of higher education come from private institutions, while in California 80 percent of the graduates come from public schools. This difference highlights the potential for great regional disparities in the ability of higher education to respond to new needs, depending on how higher education is financed in different regions in the future.

It is difficult to say which sector, private or public, and therefore which states and regions, will be most affected by the ravages of inflation and the cutbacks in federal government spending. In some states, the question is being raised whether state support will go only to public institutions or whether it can be used to help beleaguered private institutions as well. How state government legislators shoulder this new strategic responsibility will impact not only the economic vitality of their state and region but that of the nation as well.

To date, the record of state government programs in support of science and technology is not impressive. In 1982, the National Governors Association released a survey of programs by twenty-four states to stimulate technological innovation.[9] The total of state funds identified by this survey came to a paltry $113,000,000 with an added $20 to 30 million for matching grant projects in two states. This sum included expenditures on new energy research, education, venture capital funding, and a variety of other programs. Three quarters was accounted for by three states, North Carolina, California, and Illinois. This sorry condition of funding for technology reflects the prevailing mood of fiscal austerity that shows little promise of expanding sources of funds.

Without the possibility of state aid, many private institutions will find themselves in especially difficult straits and even more dependent on tuition income. So far, students and parents have been partially buffered from large inflationary tuition increases by rather generous government loan programs, but as these get cut back, private schools will lose their ability to further increase revenues from tuition. Even at present tuition levels, they may already be losing an increasing number of students to the public sector. Private endowments, which used to offset the lack of state government income, have been substantially eroded due to inflation and a stagnant stock market. They now provide less than 5 percent of private school income.[10]

Edmund Cranch, president of Worcester Polytechnic Institute (WPI), notes that "the burden of higher education cost has shifted from grandfather to father and now to the student himself."[11] Kenneth Ryder, president of Northeastern University, says that "since demographic declines are most severe among upper-income families, the only students who will be able to pay full tuition won't exist!"[12] Under the circumstances, some form of a widely accessible student loan program seems to make more and more sense. To the extent that higher education permits students to get better and higher paying jobs, then students ought to be able to contribute to the cost of their schooling and sustaining the system of higher education.

The demand for student loans has mushroomed in recent years. While it is perhaps appropriate to reduce or even eliminate some loan subsidies, it would be a tragedy to withdraw in toto this considerable source of funding, not only for the student but for the nation as well. Many schemes have been proposed to create a universal student loan pool that would be paid back automatically through payroll deductions.[13] By whatever means, however, it is essential to continue to permit students to invest in their own education.

The Pros and Cons of Industry Financing

What about private corporations? To what extent can they, or should they be expected to, fill in the gaps created by federal cutbacks? Over the past four decades, giving by all U.S. corporations has been relatively flat when viewed as percentage of net income before taxes. In 1981 that sum totaled an estimated $3 billion of which only 25 percent, an amount equal to about $750 million, was devoted to higher educational purposes.[14] Yet the potential of much higher giving rates remains untapped. In 1981, despite increases in total giving over the prior year, the Council for Financial Aid to Education estimates that only 23.4 percent of the nation's 2.5 million corporations donated to nonprofit activities – educational or otherwise.[15]

UNTAPPED POTENTIAL: Corporate Contributions

	Amount (in millions of dollars)	Percent of net income before taxes	Percent of net income after taxes
1936	30	0.39%	0.61%
1940	38	0.38	0.53
1945	266	1.3	42.92
1950	252	0.5	91.01
1955	415	0.8	41.52
1960	482	0.9	71.78
1965	785	1.0	21.70
1970	797	1.0	61.93
1975	1,202	0.91	1.47
1980	2,700 .est	1.10	1.65
1981	3,000 .est	n.a.	n.a.

Source: Department of Commerce, Internal Revenue Service, Council for Financial Aid to Education (October, 1981).

There is evidence, however, that industry is responding to the new needs. Corporate gifts to higher education were up sharply in 1981. Also, many members of the American Electronics Association have been asked to pledge 2 percent of their annual R&D budgets to support engineering education, which could amount to several hundred million additional dollars per year targeted mainly at engineering education. Some companies, like Digital Equipment Corporation, already far exceed that giving rate. According to Andrew Knowles, vice president of Digital, grants to education are almost 5 percent of the annual R&D budget.

Yet while greater corporate support can have a positive impact on specific segments of higher education like engineering, it would be folly to believe that private industry can fill the gap left by federal cuts. At the current rate of $725 million, even a doubling of corporate giving would not make up for a mere 10 percent cutback in federal support. The federal government's role in higher education is significant and too important to allow for any sizable retrenchments, at least in the near term.

Where corporate support can make a difference is in selected slices of the higher education enterprise. Engineering education, for obvious reasons, is of greatest interest to high technology companies, and if corporate contributions are forthcoming, engineering education may have a comparatively brighter future. The overall cost of engineering education can be estimated to be at least $4 billion, and probably much more considering that the cost of educating an engineer is more than for other disciplines. If the high tech industries were to contribute 2 percent of their R&D budgets (an amount proposed by the American Electronics Association) to support engineering education, then an additional $400 million might become available[16] – a potential increase of up to 10 percent in engineering educational capacity. While enough to make a difference, even these amounts will need to be leveraged by government funds.

The prospect of industry financing of university programs can be controversial on campus. Recent proposals for new ventures generated a number of disagreements: at MIT over the proposed semiautonomous $125 million Whitehead Laboratories; in Connecticut over investments by Wesleyan University in private corporations (the sale of one grossed $100 million for the university); or in Boston over Monsanto's purchase of patent rights emanating from biomedical research at Harvard Medical School financed by a $23 million grant.

What these cases exemplify is the need to develop ground rules for industry/university relationships. Leading university presidents seem confident that guidelines can be worked out. President Terry Sanford of Duke University feels the question of rights and royalties is clear-cut. He says:

> We should take full advantage of commercially worthwhile activity. I do not think that we should just license out a valuable idea if we can do better in some other way. And for those who worry about where the money goes, that is a university decision. I don't even worry about distortions because the amounts in question will most likely remain small.

Sanford does draw a precise line between "having a discovery become a source of income" and having the university "invest" in a business operation. The latter, in his view, would be ill advised.[17]

At Rensselaer Polytechnic Institute (RPI), commercial arrangements at the university have gone a step further. Space and services are made available to student-founded companies in return for equity under a program entitled the Incubator Space Project. But even to beneficiaries, the distinction between academia and business is a concern. Mark Rice, president of a $300,000 a year incubator company, reflected in a press interview: "I think the university knows it is on a thin line here." Education and industry "are two distinct entities that should not be combined. . . . The outer limits of this kind of thing are not very far from where we are now."[18]

Academia's worries about greater dependence on industry funding are capsulized by Professor Everett I. Mendelsohn, a history of science scholar at Harvard. The university mandate, he states, "might then become 'Do research in areas that will lead to profitable markets, not just those which add to the body of scientific knowledge'."[19] Quoting an unnamed Harvard administrator, the *New York Times* reported the university dilemma: "What we want is to get pregnant without actually losing our virginity."

NEW DEALS FOR THE 1980s?

Harvard University: — Monsanto budgets $23 million for cancer research at Harvard Medical School and acquires all rights in precedent-setting agreement.

Harvard University: — Dupont gives $6 million to Harvard Medical School for biomedical research; *Joseph E. Seagram & Sons, Inc.* provides $5.8 for alcohol metabolism and alcoholism research.

MIT: — Exxon Research & Engineering Company to provide $8 million over ten years for fossil-fuel combustion research; controversy over smaller projects dropped because of allegations of Exxon influence.

MIT: — Whitehead Institute for Biomedical Research founded with endowment from Whitehead family; semiautonomous entity at MIT caused rifts in faculty feelings about loss of control over the research agenda and appointment of staff.

MIT: — Flow General, Inc. bought license rights for cell culture methods for a minimum annual royalty of $400,000 for the life of the agreement.

Rensselaer Polytechnic Institute (RPI): — Raster Technologies, Inc. founded by full-time students in "incubator space" within RPI Computer Graphics department. Provides RPI a minority equity share in lieu of rent and services. Five other firms have similar arrangements.

Rensselaer Polytechnic Institute follows Stanford University example in developing a 1,200 acre industrial park and adds offer of tenant access to RPI computers, faculty consulting, venture capital counsel, and library services.

University of California: — Emerson Radio Corporation buys commercial rights to patents from computerized scanner research in return for $l.5 million contribution to the project.

Wesleyan University: — American Education Publications bought by the university in 1949 for $8.2 million and sold to Xerox in 1965 for a gross return of $100 million; in 1970 the university provided $100,000 in seed money to a high tech firm, *Zygo, Corp.*, which now has annual sales of $10 million a year.

Worcester Polytechnic Institute (WPI): — Emhart Corporation is "investing" $700,000 a year to provide a robotics laboratory and learning environment for foremen.

Yale University: — Celanese Corporation has contracted Yale for a three-year, $1.l million effort on enzymes relevant to chemical and fabric production. Yale will hold patent rights from discoveries and Celanese gets exclusive rights for an unspecified time period.

What Will It Cost?

To put the financial picture into perspective, it is tempting to speculate what it would cost to upgrade and expand engineering education. Such an exercise, however, is fraught with difficulty and peril. Solid, aggregate cost figures are not readily available, at least not in a form easily understandable to industry financial people. Costing out educational strategies turns out to be as frustrating to university and college administrators as it does to company presidents. Even assuming that the problems of education could be reversed by dollars alone — i.e., setting aside the issues of quality for a moment — it is a herculean task to get academic accounting systems to yield answers to the seemingly simplest cost questions.

DEALING WITH ELUSIVE COSTS

Imagine two presidents alone in a corporate conference room trying to strike a deal. One sports a gold watch cum calculator, the other smokes a pipe. The calculator wearer runs a large, growing computer company with 5,000 employees and plenty of cash. The pipe smoker runs a large, urban university with 15,000 students and plenty of bills. The industry president needs more scientists; the university president needs more teaching staff. It comes time to discuss the bottom line:

"How much do you need to support an additional engineering graduate?", asks the corporate president, his calculator at the ready.

The pipe lights up. "Well, doctorates are twenty times more than masters, and masters five times more than bachelors. Which do you want?"

"Let's try masters."

"With thesis or without? Our faculty insists on the thesis for quality reasons, but it takes 25 percent more time and money."

"All right, but how much?" The calculator gets impatient.

The pipe smoke thickens. "Well, if it's a computer science grad, he'll need a computer as part of his studies. Then there's lab equipment, not to mention renovation of the floor space required. At the moment, we've been unable to fill several vacancies on our computer science faculty, so I'll throw in the cost of a new professor. Besides computing, do you want your graduate to read and write?"

"Read and write. . . . ?"

"Yes, we'll have to prorate part of the remedial training and some of the English lit department into the cost. What about sleeping? The dormitories are presently at capacity, so there'll be some construction costs to be factored in."

"Hey, I didn't ask to endow a dormitory! All I want is a good man who knows computers."

"A man?" puffs the pipe. "We ran out of them a long time ago. But for 20 percent more, we can boost our high school program to get more women into engineering."

Despite these uncertainties, some of the more important cost parameters can be indicated here. Most engineering department heads agree that increasing support for their faculty is the primary problem in upgrading engineering education. With surging undergraduate enrollments and difficulties in retaining qualified faculty, student-faculty ratios are way out of line by historic standards — 40 percent more onerous than ten years ago. Thus the most frequently cited needs are (1) for increases in faculty salaries to make them more comparable with industry; (2) for provision of graduate student fellowships, which are needed to support the masters and doctorates who eventually replenish the faculty supply; and (3) funds for up-to-date laboratory equipment for research and training. Let's look at what might be required in each area.

Faculty salaries:[20] Calculated on the basis of 1979-1980 rates and an academic year of nine months, salaries range from $45,000 for a department head to $12,000 for an instructor. The national averages are $28,000 for an engineering department head; $26,500 for a tenured professor; $21,000 for an associate professor; $18,000 for an assistant professor; and $14,000 for an instructor. The overall average for a full-time faculty employee is $25,000 – which is about the same figure that the professor's student will be offered as a starting salary from industry. On average, faculty salaries range from 20 to 30 percent below industry rates.[21]

In 1980, there were approximately 20,000 budgeted full-time instructors/professors of engineering. In addition, there are some 1,600 unfilled job openings for full-time engineering faculty, or 10 percent of the total pool. The cost of filling 1,600 vacant positions would be $40 million at an average salary of $25,000. In addition, the cost of a 30 percent increase in salaries to make academic employment fully competitive with industry would amount to $150 million if the entire national engineering faculty were "given a raise." For comparison, this is an amount equivalent to six and half hours of expenditures by the Department of Defense.

Graduate fellowships: Probably one of the greatest deterrents to becoming a professor is the long period of graduate studies. During this time, income is only about one-third of what the same student could earn in industry. The graph below shows the case for MIT. Over the past decade, the median monthly starting salary in industry has risen from $900 to $2000, but teaching assistant fellowships have gone from $300 to only $750, putting the university more than ten years behind industry.

Extending financial support to graduate students at the master's and doctoral levels might entail offering average annual support stipends of approximately $10,000 per year per student. If this were made available to, say, half the graduates enrolled in engineering, or about 10,000 students, the cost would come to $100 million.

Equipment update: If the nation embarked on a crash program to update *all* equipment resources in U.S. engineering departments, Daniel Drucker, dean of engineering at the University of Illinois, estimates the full cost at $1 billion.[22] To put this on an annual basis as the other numbers are, the useful life of the equipment might be estimated at seven years, thus yielding a per year figure of $140 million. (Drucker also estimates that there is a $1 billion need for new buildings. Prorated on a ten year basis, this would comprise another $100 million.)

MIT B.S. ENGINEERING GRADUATES:
Starting Monthly Salaries

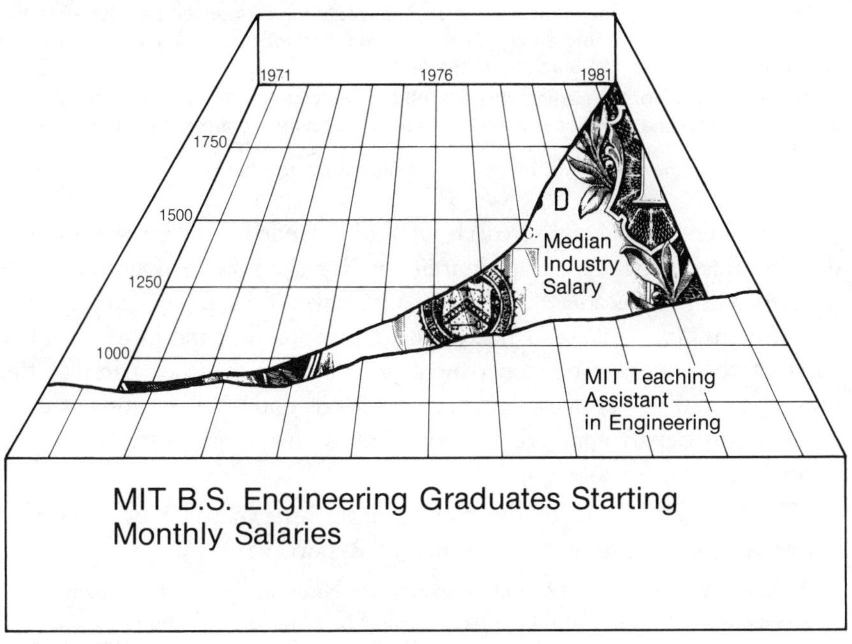

Source: MIT Placement Bureau

Considering the leverage of engineering education on the economy, these estimated numbers are not excessive. And when national planners can afford to talk about a five-year $1,600 billion defense budget, they should not be unattainable.

The figures estimated above were for all of engineering education and, if made available, would go a long way toward reversing the declines already under way. Many high tech executives would like to know what it would cost to substantially expand the output of graduates in one particular slice – especially in electrical engineering and computer sciences, which represent about one quarter of all engineering graduates.

A NEW DESIGN AUTOMATION LAB:
What It Cost in Minnesota

The University of Minnesota decided in 1980 to upgrade its computer-aided design laboratory. The "bare bones" cost included $600,000 for a time-sharing graphics computer (donated by the Calma Company), $140,000 for facilities preparation, and an annual laboratory maintenance and operating cost of $90,000. Due to space limitations, the laboratory could not be housed in a single building but had to be scattered among a variety of unconnected locations making access for students inconvenient. However, the news of new equipment and an expanded program had a dramatic effect and boosted enrollment from 25 to 125 students in the first year.

Interestingly, 63 of the participants are full-time university students while the remaining 62 are part-time enrollees for whom industry provides double the normal tuition costs.

A great deal of careful research would be needed to estimate with any degree of accuracy the cash requirements for example to double EE/CS degrees within a decade.[23] Such an estimate would be subject to arbitrary assumptions. It would also necessitate corresponding increases in other parts of the engineering department, and indeed in other parts of the university and its services. Further, the needs and opportunities of each engineering department are too different to make any such aggregate figure readily estimable.

One recent estimate of the cost to outfit an expansion of university researchers was made in the "Snowbird Report":[24]

> Assuming an average Ph.D. student spends four years in the Ph.D. program, an average of 1,000 must be in the current pipeline to achieve a graduation rate of 250 per year. Counting 840 Ph.D. Computer Science faculty, we estimate 1,840 university researchers. If each is capitalized at the average of $30K, the total capital investment would be $55.2M.
>
> For a different estimate, suppose that the top 200 faculty supervise 400 of the graduate students using frontier facilities capitalized at $70K per researcher; these 600 researchers require an investment of $42M. Capitalizing the remaining 1,240 researchers at the average of $30K each requires an additional $37.2M. The total investment in this case is $79.2M.
>
> The above figures show the investment for the current number of researchers. Growth of 50 percent in the number of Ph.D. students would increase the required investment to about $95M.

Another way to estimate the costs of expanding the number of EE/CS graduates is to calculate the cost of an additional faculty member. Estimates based on the Engineering Education & Accreditation Committee of the Engineer's Council for Professional Development[25] show the following:

Assume a salary of $30,000; equipment costs of $70,000 (over seven years, equal to $10,000 per year); floor and office space at an average of 1,400 square feet of space at $120 per square foot or a total of $180,000 (or $12,000 per year over fifteen years). In addition there are associated curricular costs of $27,000 per faculty. This yields a total of about $80,000 per new faculty member. University overhead rates would easily bring this figure closer to $100,000 to $125,000. At that rate, a thousand new faculty members nationwide would need a minimum of $125 million.

Yet another way to get a handle on expanding output is to look at the costs of new microelectronics centers that are springing up in several universities. Some of these were described in considerable detail in the prior chapter, because they represent an exciting new approach to industry/university partnerships. Here we highlight some of the cost considerations.

Building costs of a microelectronics research and teaching facility generally are in the $10 to $20 million price range. Annual operating budgets run $3 million or more per year. Stanford University's new VLSI microelectronics center will be housed in a $12 million facility. Just the rennovation of MIT's new VLSI facility will cost $10 million. In North Carolina, a similar effort under way at the Research Triangle Park has a start up cost of $24 million, two thirds in construction capital and the balance in operating funds. Additional research efforts to be sponsored by the Semiconductor Research Cooperative will divide a budget of $20 to $50 million among three to six centers to support the expenses of research in microelectronics. In 1982 in Massachusetts, legislators gave preliminary approval to a bill requesting $20 million, to be matched by an equal amount by industry, to build a semiconductor processing laboratory and teaching center managed by a consortium of eight universities.

The purpose of this accounting exercise is not to produce a hard-and-fast number but rather to get an order-of-magnitude feeling for the cost implications of the engineering problem. Considering the leverage involved for the economy, the numbers involved are not excessively large in relation to the total investment of $65 billion in higher education, and are dwarfed by the projected expenditures of more than $200 billion currently earmarked for defense. The message is that targeted strategic investments in education can have a big payoff for the economy.

What Education Can Do

Beyond the fundamental question of money and budgets, educators are proving adaptable in responding to the new needs of a knowledge-intensive society. Two interlinked areas with great promise in this regard are lifelong learning and the use of new educational technologies, especially instructional television.

Lifelong learning: The idea of lifelong learning is not new. It was used in the agricultural extension colleges introduced by the Morrill Act in the 1860s in the form of adult education for farmers. In the early 1970s, the concept was internationally endorsed by UNESCO as a goal for each country's educational system.[26] The idea has had two main parts: to make education available to adults throughout their lives (variously called adult education, continuing education, or recurrent education) and to refashion the traditional school system to reflect the fact that people will come in and out of "school" throughout their lives. These two ideas combined comprise the full meaning of lifelong learning (or what the French call "education permanente").

While a good idea in theory, lifelong learning has only recently been considered a serious possibility. For many reasons, the adult education market is exploding in the United States as well as in other developed countries. At MIT, there is a long tradition of educational outreach to continuing students, not just in engineering, but in many forms of technology education. The Center for Advanced Engineering Study, directed by Myron Tribus and his deputy Jack Newcomb, has actively pursued a variety of programs in lifelong learning. In electrical engineering and computer sciences, a new proposal is now being considered by Professors Fano and Smullin to establish an expanded program in lifelong cooperative education.

Electrical engineering and computer sciences have an estimated half-life of five years. This means that half of a computer scientist's knowledge is obsolete in five years. Since it takes about five years to acquire the knowledge, the EE/CS graduate is always running to catch up. The MIT program, and others like it, would provide a way to keep out in front of new knowledge.

Lifelong Learning at MIT

"Is the pace of change making you dizzy? Do you wonder what the bright young graduates are talking about?"

So began a recent letter to MIT alumni inviting them to take part in the 1982 centennial celebration of electrical engineering education at MIT. As part of the 100 year's celebration, Professor Robert Fano was asked to chair a committee to recommend what MIT's department of electrical engineering and computer science should do in the area of lifelong learning, especially in electrical engineering and computer sciences, where a major new development every four to five years puts a premium on keeping up to date.

Several programs are under discussion in Fano's group. One would be an off-campus program for junior engineers, say those under thirty. By spending at least one term on campus and taking four to six courses off campus over two to three years, they could earn a master's degree while still being employed. Possibly anywhere from 150 to 400 new students could be enrolled in such a program.

Another program would be for senior engineers, say those over thirty. Here it is unlikely that a degree would be awarded, although some certification could be considered. What is most needed to make lifelong learning an institutional reality, the MIT committee says, is to create an environment conducive to learning at the company site. This means: developing a "distributed faculty" — lecturers at the company who become part of an MIT extended academic community; creating new courses, for example in robotics, VLSI design, knowledge-based systems, and computing of very complex problems; and finding new sources of financing.

Professor Fano is well aware of the pitfalls that can beset any new proposals for change. "We are shooting high, but we have to start low," he says. He resists the idea of spelling out what institutional commitments would be required to make lifelong learning a reality at MIT, but one is clear: industry will have to foot the bill. Given Fano's interest in forming true partnerships with industry to develop the new materials, there should be a lot of company support. Interestingly, the first group who will test any new course materials are likely to be MIT professors themselves, who are as much in need of updating outside their own areas of expertise as are industry engineers.

If what we want are more senior engineers, we should turn more to those who already exist. Increasing the supply of new engineering graduates — even if it occurs quickly, which it most likely will not — will contribute to advanced or senior levels only after seven to ten years. In the meantime, many senior engineers are exiting the profession for management and other career paths that are more rewarding. The lack of professional identity and corporate clout, plus continual pressure from recent graduates, often leads to a mid-life crisis in engineering that is serious for the individual and the company and exacerbates the national engineering shortage.

It seems that no one wants to be a fifty-year old engineer in America, but America needs them all desperately. Engineering is presently less attractive than the medical or legal professions as a lifelong career, which is reflected in higher income levels for doctors and lawyers later in life. An important goal of lifelong learning should be to elevate the professional status of engineers in industry to increase the viability of engineering as a lifelong career.

Technical obsolescence is certainly one of the problems that older engineers face as bright young graduates come out of universities steeped in the latest technology. Obsolesence is exacerbated by the fact that the complexity of modern technology increasingly obliges individuals to specialize narrowly to be successful in their organization. When a new wave of technology breaks, which occurs now with increasing frequency, the worth of an engineer's prior experience is substantially devalued, or at least it is perceived that way. In reality, the mature judgment, depth of knowledge, and organizational experience of mature engineers are invalaubale assets. A greater commitment to lifelong learning by companies and by engineers themselves can circumvent or at least greatly ameliorate the issue of technical obsolescence.

But there are other more subtle issues that detract from engineering as a lifelong career. In many companies, senior technologists do not have sufficient power in the organizational structure to influence corporate policy and strategic decisions commensurate with their abilities. This leads to frustration and is another reason to opt for a management rather than a technical career.

These interrelated problems – inadequate financial rewards, accelerating technical obsolesence, and insufficient involvement in the corporate decision process – will have to be solved to conserve the technical resources already in place. There are reasons to be optimistic that technology itself will play an important part of the solution. Already, sophisticated computer-aided design and automatic test equipment are reducing the routine drudgery of engineering tasks. This makes technologists more productive and valuable and their jobs more interesting and exciting. Equally important, technology developments are revolutionizing educational delivery systems and making continuing education of working engineers more practical and feasible.

New Modes of Education

Paralleling trends toward lifelong learning are new developments in educational technology. Both are helping to solve the engineering education capacity issue, as well as helping each other. For example, one problem in lifelong learning is logistics. If the student's place of work is far from the university, much is lost commuting. Transporting the classroom to the workplace via instructional television (ITV) overcomes this obstacle.

There are two main technologies for education at present: computer-based instruction and ITV. Eventually, they are likely to merge, especially as telecommunications advance. But even before they merge, each can contribute to engineering education.

Computer-Based Instruction: One of the most important new tools for teaching engineering is computer-assisted design (CAD) equipment. This gives computer assistance to electrical engineering students to design and test circuits that are too complex to do by hand. The addition of CAD equipment to the University of Minnesota increased course enrollment from 25 to 125. At Duke University, new CAD instructional equipment will be a major part of Duke's participation in a new North Carolina Microelectronics Center.

The use of computers in education is undergoing a rebirth as microprocessors and personal computers become more prevalent. The large central data processors of the late 1960s proved not to be conducive to what was then called computer-assisted instruction (CAI). But as of the early 1980s, the experiments with new approaches to education via mini and microcomputers is extensive and promising.[27] The U.S. Department of Education has sponsored a series of seminars on the uses of microcomputers and videodisks in 1982, and the U.S. and Canadian delegations to UNESCO have stressed education and microelectronics as their top priority for the next three to five years.

Instructional Television (ITV): While computers are likely to have a major effect on EE/CS education, instructional television is already making its mark. This trend is tangibly evident to the west of Palo Alto, California, on a low range of undeveloped hills stretching north-south. They form an alluring skyline of dry grass ranges, gnarled old oak trees, and aging eucalyptus groves nestled in the low areas that border the roads. Lying wide and flat further to the east is Santa Clara Valley, better known as Silicon Valley, but the only hint of its intense technological activity in the Palo Alto hills is a broadcasting tower with a cluster of dishes on Black Mountain. These innocuous transmitters, five miles from the sprawling Stanford University campus, are the lifeline between the university's Instructional Television Network and industry audiences in Silicon Valley.

In an amphitheatred engineering classroom on the Stanford University campus, a course instructor emphasizes a point to his attentive class. Suddenly from a loudspeaker, a voice pipes in:

"Professor, may I ask a question?"

"Go ahead," the professor responds to the faceless voice.

The questioner was three miles away watching the class on a television screen at a Hewlett-Packard plant. He was one of several thousand full-time employees in Silicon Valley taking a Stanford University engineering course without ever leaving his place of work.

One hundred companies in the Valley receive, live, Stanford's broadcasts of engineering and science courses. Up to four classes can be broadcast simultaneously with the added advantage of two-way audio contact between the students and professors on the main campus. Many of the several thousand participants, who are full-time industry employees, are part-time candidates for master's degrees through an Honors Cooperative Program offered by the departments of Engineering or Computer Science. In addition, videotaped class cassettes are offered to thirty other locations beyond the broadcast range of the network on the condition that companies provide a tutor approved by the course instructor.

The productivity increases and the broadened reach of the teaching faculty are reflected in the financial success of the program. It generates a net yearly revenue of $1 million above the basic tuition fees. This income is earmarked for use at the discretion of the engineering department. Part of this is distributed to faculty members as unrestricted funds for research — a highly valued perk at a time of budgetary restraint.

While outstanding in the coverage and financial success of live television course offerings, Stanford is hardly unique in the use of broadcast technology. Teaching by instructional television is routinely offered in various degrees of sophistication by about thirty universities and colleges nationwide. These reach an estimated 44,000 engineers at their places of employment. A vital feature of this burgeoning television medium for teaching, according to the National Academy of Engineering, is that "these ITV systems are local responses to a professional need and are paid for by industrial and government employers. . . . No Federal funds are involved."[28]

Some of the better known live broadcast programs are those at the Illinois Institute of Technology and the University of Southern California, as well as those of the innovative university and college consortium linking nine institutions in north Texas under the innovative Association for Higher Education (known earlier as the TAGER Television Network).

In January 1982, the Institute of Electrical and Electronics Enginners inaugurated live broadcasts of engineering relayed from South Carolina by satellite to thirty-eight other sites in the United States. An even more ambitious undertaking that would lead to the creation of a National Technological University is being investigated. It would broadcast an accredited master's program based on coursework offered by a consortium of member universities. Computer engineering is considered a priority program for the proposed university as it is envisioned by the Association of Media-Based Continuing Education for Engineers. Founded in 1976, the AMCEE is located in Atlanta, Georgia and is a cooperative effort of twenty engineering universities.[29]

An alternative to live broadcasting — noninteractive videotapes of classes — has been steadily cultivated into a major teaching resource by MIT. The university stands out nationally with an inventory of cassette tapes totaling more than 2,000 classes in sciences and engineering. MIT's Center for Advanced Engineering Study (CAES) pioneered the development of continuing education packages. These modular courses consist of a coordinated set of videotaped lectures/demonstrations, study guides that include all of the visual aids used in the lecture, and computer decks and manuals. The average length of a course is twelve to fifteen lectures, each about thirty minutes long. The content is oriented to practicing engineers, not young graduate students. The program, now self-supporting, is the largest effort of its kind in the world.[30]

Television has been with us for a long time and its uses as a tool for instruction are hardly novel. What is of current interest is that instructional television has matured in Silicon Valley to become a highly effective device for live teaching broadcasts of engineering courses. Much of this evolution flows from ten years of experience following a crucial decision in 1971 by the Federal Communications Commission to reserve twenty-eight six-megahertz channels for nonprofit educational purposes – and which may, under a new FCC ruling, be reverted back to commercial use for what is called multipoint distribution. The channels were labeled in the early 1960s as Instructional Television Fixed Services (ITFS). In the words of an observer, John A. Curtis, "today, for the first time in history, an instructor is able to 1) teach students in one or more geographically remote classrooms on the basis of direct face-to-face relationships; 2) deliver 'written' materials to remote locations from a central point; and 3) have low-cost digital circuits available to access, control and distribute remote computational power and data bank information."[31]

An additional feature that makes the technology stand out as especially important at a time of faculty shortages and high training costs is the cost effectiveness of instructional television. On a per-student basis, the average annual operating costs for graduate-level systems were computed in 1979 to be about $35 per student for the capital investment in equipment, and about $80 per student in operating costs.

Past studies have demonstrated that "both math and science courses are equally well taught face-to-face or by television." With the added presence of a tutor providing an opportunity for personal contact, the quality of education provided by television was impossible to differentiate from classroom instruction delivered in traditional fashion.[32]

The use of communications technology as a tool for expanding the productivity of educators is apparent, but its application remains controversial. In many engineering departments, including Stanford's, resistance to the use of broadcast classes is strong. Two fears stand out in the minds of many faculty. The first is that the rapid rise in numbers of 'remote' students will pressure faculty to spend more and more time monitoring examinations and attending to the associated administrative work. When it comes to extending the participation of industry-based students in the degree granting master's program, the resistance is even stronger. In the words of Engineering Dean William Kays, the Stanford faculty resistance may have more to do with tradition than substance. "I think by nature," he states, "faculty are suspicious and resistant to 'volume'."[33] They worry about the dilution of the Stanford name and reputation as courses are broadcast beyond the immediate and manageable boundaries of Silicon Valley. The fear is that at some invisible point in time or in scale, the program no longer *is* Stanford.

On the other hand, the new technologies have the potential to augment the prestige and income of teachers who successfully master them. MIT's tape on "Digital Signal Processing," authored by Alan V. Oppenheim, a professor of electrical engineering, and produced by MIT's Center for Advanced Engineering Study, is a best seller. Over thirty thousand students worldwide, primarily practicing engineers and technical managers, have made this the most widely used nonbroadcast videocourse in the world. Since its inception in 1975, the tape has grossed over $1 million for MIT. Professor Oppenheim himself has become something of a celebrity in his field. His textbook, also entitled "Digital Signal Processing," is one of publisher Prentice-Hall's most popular in the college textbook division. Oppenheim receives royalties not only on the textbook, which is the traditional practice, but also on the videotape, which is the way it should be to encourage more faculty members to expand their work.

Instructional television for engineering and computer science education represents an immediate remedy ready and waiting for development. Stanford and MIT provide two successful models, one live and one pretaped, of television-based methods to extend the capacity of traditional classroom education.

Introducing more technology into education, changing traditional delivery systems toward lifelong learning, entering into new partnerships with industry and government — all these issues and others raise fundamental questions about broad educational policy. What are appropriate goals for higher education in the twenty-first century? What do students entering a knowledge-intensive era need to learn?

These important policy issues are presently tied up in disagreement among professional educators. This controversy focuses on the proper role of the humanities and the liberal arts at a time of general cutbacks and strong pressures toward more technology education. To what degree should educational goals focus more on education for employment? Will more money for technology education necessarily translate into less money for the humanities? We turn now to these issues, known as "the humanist-technologist" debate, and offer some thoughts on directions for a "new education" appropriate to knowledge-intensive society.

Chapter 8

THE NEW EDUCATION

IN EARLY 1982, the Andrew W. Mellon Foundation pledged a $24 million investment for humanities students in history, art, musicology, philosophy, and religion. Some months earlier, the Exxon Education Foundation announced a $15 million program to provide 100 teaching fellowships in chemical engineering, earth sciences, and other engineering fields, including electrical.

What is remarkable about these two independently conceived initiatives is not their obvious differences in purpose – one focusing on classics, the other on engineering – but rather their not so obvious similarity of concern with the employment market. Humanities-related jobs are scarce. "This threatens the vitality of humanistic studies into the twenty-first century," according to Robert Goheen, president of the Mellon Foundation.[1] The conclusion: more support targeted toward academic employment is needed precisely because the present job market for history and classics is so bleak – less than half the number of jobs are projected from 1981 to 1995 than were available between 1971 and 1975.

Technology-based jobs are abundant. This threatens the viability of teaching in science, technology, engineering, and math, as many teachers leave the profession for better paying jobs in industry. Therefore, the purpose of the Exxon grant is to encourage more technologists to take up teaching as a career.

Both the Exxon and Mellon grants represent responses to the increasing mismatch between older educational ideals for enlightenment versus the shifting requirements of a new economy. Both also underscore another critical point: that educators are severely underfunded at a time when strategic changes need to be made both in the content and structure of education. Bringing today's educational system up to speed for a knowledge-intensive society is not going to be easy. Much needs to be done.

Science and technology education must be strengthened considerably — on a par with efforts being expended by competing nations. Liberal arts programs need a stronger grounding in technical and quantitative subjects to prepare students for employment in a technology-oriented society. Engineering and science programs need broader exposure to humanities to strengthen their conceptual, communications, and leadership skills. We need more Science, Technology, and Society programs (STS), which interrelate the viewpoints and values of humanists and technologists.

Universities must become more responsive to the needs of society and of students, and more flexible in the way education is provided. Four years is no longer adequate to develop the intellectual capacity of students while giving them specific knowledge and skills required to begin a meaningful career. Continuous education toward advanced degrees in tandem with employment will become much more essential in the future. This will require innovations in off-campus education by instructional television (ITV) and other advances in educational technology. The productivity of faculty must be increased through videotape and computer-based technologies with paraprofessional support such that teaching will remain a financially attractive and viable career.

All this requires money. Some new money can come from high tech corporations, particularly as safeguards are developed to ensure academic independence. Some will come from adult students in on-the-job continuing education. But government, both state and federal, will have to sustain the high level of support that has characterized the system of higher education in America.

Strengthening Technology Education

Pressures are increasing to shift the educational mandate toward preparing students to get better jobs. The shift already under way is clear in the data that shows an increase in professional degrees such as business and law and a decline in the number of liberal arts degrees such as English and history.

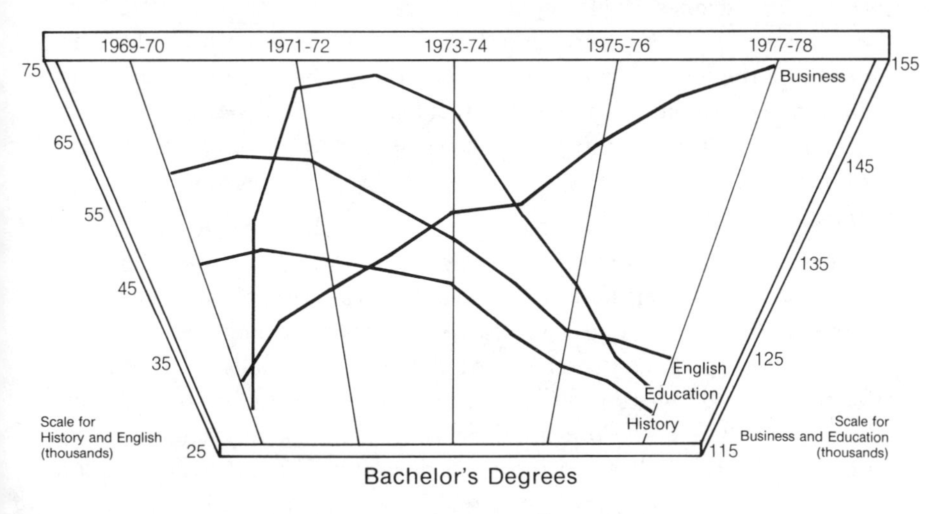

The Decline in the Liberal Arts

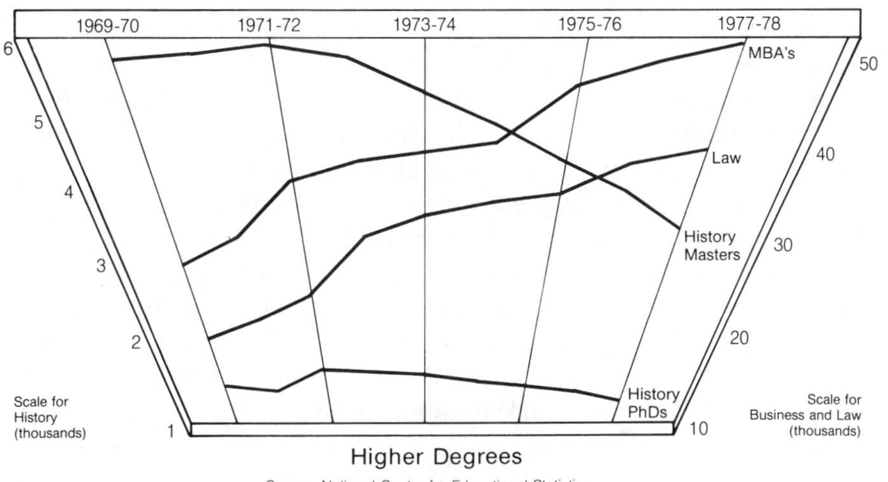

Source: National Center for Educational Statistics

Technology-based education needs to be strengthened, especially in engineering. The number of degrees in engineering has remained more or less flat in the last decade. "We've had an easy slide,"[2] says Lester

Thurow. Since the flurry of scientific education activity surrounding Sputnik, students have been avoiding the path of hard sciences. The lack of introductory science in elementary schools, the declining math requirements for a high school degree and the falling high school math and science scores, and the lower percentage of students in engineering-based higher education (6 percent in the United States versus 37 percent in West Germany and 21 percent in Japan) are testimony to the slippage in technical education.

The causes of the easy slide cannot be pinned solely on teachers, administrators, or other educators. Society, from parents to government agencies to students, has participated in the steady deterioration of our educational systems by reducing the pressure to work hard at learning and by lowering expectations of competence and demonstrable skills. It is time to reverse the trend toward "soft" education, and to give full support to those educators ready to better prepare their students for the future. This means harder work, higher levels of personal development, and greater educational achievement – in arithmetic and also in reading and writing.

The quality and quantity of the training for science and technology graduates has to be increased. This is true not only for greater technological competence, but for broader purposes as well. Corporate managers complain that engineers literally cannot read and write, and that they are too narrow in their conceptual skills. And not all engineering graduates will devote themselves entirely to engineering careers. While many end up designing products, an even larger number will become managers in sales, manufacturing, and marketing. An increasing number will become general managers, presidents, and entrepreneurs. It would be desirable if a greater number would become congressmen, senators, and governors. Engineering and science, as well as liberal arts, can and should be a valid starting point for diverse careers, but this will require greater development of intellectual capabilities and exposure to broader thinking about the world at large.

The United States is becoming a rapidly changing technological society. If science education leads toward technological competence, humanistic education is needed to provide leadership in an era of rapid change. A recent survey of college students studied their preferences for education – to get a better job or to get a better liberal education? The results show that students consider both important.

The type of education required in the coming decades should remove the limits to learning imposed by educational policies that are either too vocationally narrow or too theoretically abstract.[3] This will require balancing the goals of employment and enlightenment, and integrating humanities with technology education. To do so will require a reorientation of both technical and liberal arts education.

WHAT STUDENTS WANT FROM COLLEGE

Important Reasons in Deciding to Go to College	Percent Answering Very Useful
Get a better job	77.1%
Learn more about things	74.6
Gain a general education	66.7
Make more money	63.4
Meet new and interesting people	56.2
Prepare for graduate school	46.0
Improve reading-study skills	39.3
Become a cultured person	34.4
Parents wanted me to go	31.5
Get away from home	8.9
Could not find a job	5.8
Nothing better to do	2.1

Source: A.W. Astin, M.R. King, G.T. Richardson, *The American Freshman: National Norms for Fall 1980*, Los Angeles: University of California, 1981.

Linking Humanities and Science

Humanities education needs to concentrate more on the employability of graduates without compromising the traditional goals of intellectual development. A knowledge intensive economy requires that an increasing percentage of our workforce have professional knowledge. This cannot be left solely to on-the-job training but must be part of the formal education process, interwoven with the broader goals of higher education. For example, computer literacy, quantitative skills, and general knowledge of scientific principles are just as important to liberal arts graduates as humanistic values are to engineers. And technically oriented people need – even if they don't always want – a broader education to stimulate their intellectual development and to sharpen their communications and conceptual skills. In fact, engineering schools have gone further in requiring humanities into their curricula than liberal arts schools have gone in incorporating computers and science into theirs.

At the level of professors, teachers, and educational administrators, closer cooperation is needed between those who consider themselves humanists and those who identify with technologists. Students would be among the first to applaud greater efforts at integrating the two. Unfortunately, in today's environment, the opposite is happening. The chasm between humanist and technologist is growing wider as each battles for a larger slice of a dwindling pie. Schools are organized such that the professional interests of academicians interfere with an objective assessment of students' needs and desires for both humanistic and scientific education. This is a dilemma confronting degree-granting authorities because the times increasingly require a mix of disciplines and views.

Stanford University has put a cap on the growth of its engineering department at 1,800 students. Without such limits, President Donald Kennedy fears the humanities would be undersupported and technology education overemphasized. But can the changing mix in the job market be ignored? Does a larger engineering department necessarily mean a weaker humanities department? Can both not contribute to the development of the student irrespective of the career path selected?

George Drake, president of Grinnell College, claims that liberal education is "in danger of obliteration as technical and vocational studies expand to meet the demands of the marketplace. Engineering," he says, "is a complex and expensive discipline which could deflect colleges from traditional liberal studies while straining their resources to a point where what these colleges do best would be undermined."[4]

These fears among liberal arts educators are real enough, but how realistic is their response? Liberal arts educators need to deal more creatively with a double bind – declining demographics, which are predicted to cut high school graduates 26 percent by 1990, and increasing professionalism, which is generating pressure for more enrollments in engineering, business, and other professional fields. Of the almost one million bachelor degrees awarded annually in America, 6 percent are presently in engineering. If this percentage grows, as it is likely to do, humanities educators need to find a more positive path than the one of resistance. The most promising is lifelong learning, particularly as we increasingly realize that four consecutive years is not adequate to intellectually develop our students and at the same time prepare them for a professional career.

It is hard to imagine that many of the best centers of excellence in liberal arts education will not continue to be the important element of the educational mix they are today. Yet, it is equally hard to imagine that many other schools will not be forced to adapt to changing market needs. What is needed is more cooperation in developing joint degree programs between liberal arts and professional programs. This will require a student's time horizon to stretch beyond the traditional four years. And implementing a structural change of this magnitude will require many more years — but now is the time to start.

In the meantime, the intense emotion in the debates between humanist and technologist obscures the fact that there are at least two different issues at stake. One concerns power, another concerns values. The first is a two cultures problem, which, in the absence of sustained and far-reaching countermeasures, is likely to persist in a knowledge-intensive society. The second as a problem of where science is taking us. This is a more immediate and perhaps more tractable problem — and if not solved in certain areas important to survival, all other problems may become "academic."

TWO CULTURES

At the 1959 Rede lectures, the English philosopher C.P. Snow delivered his now-famous treatise, *The Two Cultures and the Scientific Revolution.* Skillfully combining British accuracy with cool understatement, he deplored the fact that academia — and consequently much of society — had separated into two mutually incomprehensible camps. The only thing rarer than finding a scientist knowledgeable about the works of Shakespeare was finding a "literary intellectual" who knew the first thing about Newton's Second Law of Thermodynamics. Yet the Second Law is as basic to the sciences as Shakespeare is to the English-speaking humanities.

Snow's thesis generated considerable debate and controversy. Most disputed was the implication that his two cultures dichotomy was universal, whereas in fact it may be a peculiarity of Anglo-Saxon cultures. Nonetheless, publication in 1963 of a second "Two Cultures" treatise seemed to reinforce a widespread belief in the English-speaking world that technology and the humanities had grown mutually incomprehensible to one another.

But mutual incomprehension may be only half the two-cultures problem. Another problem intensified by the competition for scarce economic resources has become mutual antagonism. The intensity of the debate has grown. The struggle now is not only for more understanding but for more power.[5]

WHERE IS SCIENCE TAKING US?

In 1981, Gerald Holton, professor of physics and history of science at Harvard University and visiting professor at the College of Science, Technology, and Society at MIT, was awarded America's highest honor for intellectual achievement in the humanities. He was chosen by the National Endowment for the Humanities to deliver the Jefferson Lectures, which he devoted to the topic "Where Is Science Taking Us?" Concerning society's view of science, he notes:

"A generation ago, the more educated the individual, the more science was approved. Distrust was largest among the least educated. Now there is a complete reversal: on the whole, the more naive, the less distrust of science."

In his lectures, Holton describes two kinds of polarization. One pits those cheered by greater advances in science against those depressed by the alienation it engenders. A second poses science's drive for autonomy against society's claim that this autonomy must be limited by social goals. The issue today, according to Holton, is "to forge an *alliance* between the business of the scientists and the business of the rest of mankind. . . ."

Recalling an earlier NEH report, Holton stressed that "*if the interdependence of science and the humanities were more generally understood, men would be more likely to become masters of their technology, and not its unthinking servants.*"[6]

The Two Cultures Problem

Tracing the politics of intellectual power back to their European roots, Princeton professor Fritz Machlup[7] shows how, in the Middle Ages, humanists were concerned not with technology but with theology. At that time, it was religion, not science, that provided the challenge. Were we living in the seventeenth century, our discussion most likely would be focusing on the humanist/theologist debate, and whether proposals from the church could overcome gaps in these two cultures. This kind of intellectual competition has existed for centuries and is not likely to dissipate soon.

There are at least four reasons why humanists are not only ignorant of the sciences but resentful of them as well. Scientists are on the winning side in a battle for money, prestige, and good students. Humanists resent their loss of dominance over the curriculum, which they held for centuries and well into the beginning of the twentieth century. Humanists have been relegated to second-class or even third-class academic citizens as natural scientists have excluded the humanities from the category of scientific knowledge, scientific research, and scientific method. And humanists reject "reductivism," which most natural scientists believe in religiously.

Thus, a main source of the humanists' antagonism toward technology is their feeling of being under attack from the scientific establishment who, in the humanists' eyes, assume an attitude of "scientific supermen."

Another aspect of the two-cultures problem is the increasing fragmentation or specialization of knowledge. Studies by C.P. Snow and other authorities like Clark Kerr (the former chancellor of the University of California) and Jacques Barzun (provost of Columbia University and perhaps America's sharpest educational philosopher), have documented the trends toward increasing "fragmentation of the university community into independent special interest groups."[8] Graduates seem to know more and more about less and less.

Solving the two cultures problem – reducing mutual antagonism as well as fragmentation and mutual incomprehension between humanists and technologists – will be a monumental, long-term task whose resolution is beyond the means and scope of industry. Wherever possible, industrial support should seek to narrow the gap between the two cultures. The president of Exxon's research and engineering, E.E. David, Jr., believes in the need to "sustain balanced, liberal education, and not have the universities changed into trade schools. Hopefully, some combination of philanthropy and enlightened industrial support will prevent this from happening."

It is important to keep in mind that it is academic professionals who are most involved in the controversy between humanist and technologist. Most students, parents, employers, and the general public decry the fragmentation of knowledge and the tendency to split into two competing camps. The challenge is for professional educators to focus more of their interests and attention on student needs than on preserving traditional channels of educational delivery.

Where Is Science Taking Us?

Increasing numbers of people are concerned that science is beyond the control of ordinary political processes and is taking us someplace we may not want to be. As scientific knowledge becomes more and more advanced, fewer and fewer people are privy to its meaning and able to control the power it confers. In the 1960s, this expressed itself as a reaction against the military-industrial complex. Today the question is more compelling: can anyone comprehend the incredible complexity of modern weapons systems, not to speak of knowing how to operate them? It seems the higher the weapons complexity, the lower the human IQ to operate them. The army now publishes its M-1 rifle assembly manual in comic book form due to low literacy rates.

In his 1981 Jefferson lectures, Gerald Holton[9] points out that this problem of complexity pervades the political decisionmaking process. Many national leaders find it increasingly difficult to make decisions on a growing number of technology-intensive projects. By one recent estimate, nearly half the bills before the U.S. Congress have a substantial science or technology component. Yet in the 1980 Congress, only 2 of its 535 members had completed any engineering training.

Pressures are increasing to have scientific advance guided by social goals. MIT now requires that 20 percent of an engineering student's courses be in humanities and social sciences. Elting E. Morison, professor emeritus of MIT, argues that every educated person should become engaged during their study in the future problems of society, which "all have their origins in technology – genetic engineering, nuclear reactors, the shuttle, and the computer, for instance." And they all spill over into social values – ethical issues of new forms of life, protests over nuclear power, military versus civilian uses of space, and the employment effects of electronics.[10] Programs to incorporate social values into scientific study are known as Science, Technology, and Society (STS) courses, and many colleges and universities have incorporated these into their curricula.

Examples of STS Courses

- There are an estimated 200 STS courses at U.S. colleges and universities. Most of these introduce humanities to scientists. Only 10 to 12 are specifically designed to familiarize the humanist with science.
- At Worcester Polytechnic Institute, the "WPI Plan" seeks to interrelate science, technology, the humanities, and arts through courses such as "Light and Vision."
- At San Francisco State University, the "NEXA" program considers the convergence of the natural sciences, social sciences, humanities, and arts.
- At MIT, the "Program in Science, Technology, and Society" offers a course on "Computer Cultures, Computation, and the Individual."
- At Carnegie-Mellon University, the "Program in Technology and the Humanities" includes graduate as well as undergraduate courses.
- In the "Program on Humanistic Studies in Engineering," developed at Princeton University in the early 1970s, students from both engineering and the liberal arts examine the social and aesthetic criteria for design in structural engineering.

- At Stanford University, the "Program in Values, Technology, and Society" proceeds from the principle that "it is vitally important to obtain a broad understanding of technology and of its human and social implications."
- CUTHA (The Council for the Understanding of Technology in Human Affairs) has sponsored conferences on technology and values at MIT and Chatham College.

An example of a highly successful program to integrate values into engineering can be found at Worcester Polytechnic Institute (WPI), a small, innovative school in central Massachusetts. In a program known as "The WPI Plan," faculty members send out teams of students to study firsthand the social impacts of technology. Former WPI president, George Hazzard, inaugurated the project because he felt that engineers had traditionally been "narrow . . . unaware of the consequences of their craft." Current President Edmund Cranch has continued to support and expand this program. Another WPI course, "Light and Vision," interrelates science, technology, the humanities, and the arts by examining the scientific, psychological, and artistic responses to light with readings from Newton's *Optiks*, Gregory's *Eye and Brain*, and Milton's *Paradise Lost.*

Most STS courses are designed to counteract the inadequacy of a narrow technical education to prepare engineers to assess their impact on complex social issues. But what about the reverse side of the coin: preparing humanists to cope with issues of increasing technological complexity? The 1980 National Academy of Engineering Task Force on Engineering Education has been looking at ways to "acquaint nonengineers with the engineering approach to problems." One school with courses such as these is the University of Central Florida (formerly, Florida Technical University). Since 1965, UCF has had requirements for nonengineers to take at least one "engineering" course. Some of the course offerings were Technology and Social Change, Engineering and Technology in History, Energy and Man, and Computers, Cybernetics, and Society. Since 1965, some 16,000 nonengineering students at UCF have taken these or similar courses.[11]

At present, federal support for STS programs is miniscule. "The whole study of the science/technology/society links, in history and today, receives federal support equivalent to the replacement cost of a couple of helicopters," according to Gerald Holton, "and even that is now supposedly being abandoned – a case of self-mutilation as irrational as smashing the headlights of your car just when you are going faster and faster into rougher and darker terrain."

It would be in the best interest of education and industry alike for industry initiated programs to encourage serious study on the question of where science is taking us. This means focusing on Holton's question rather than on Snow's. New industry/university partnerships should consider setting aside support for programs that seek to learn more about the impacts of technology on society. This could be done in many ways: by supporting specific course development; by encouraging affiliations between engineering schools and liberal arts colleges; by including more computer and technical content into humanistic programs (for example, liberal arts degrees with a computer option); by developing continuing education programs that add a humanistic content to the existing technical workforce; or by creating adult training programs to give the nonengineering portion of the population an opportunity for employment in job-abundant industries.

Asking educators to make such substantial changes when they are worried first and foremost about survival is a bit like asking flowers to bloom in the desert. Is it fair to finally point the finger at education as the key to solving America's response to international competition, human resource shortages, and overcharged defense priorities? The answer is clearly no. Most forward looking educators are willing to play their part in moving America toward a new future, but the resources to do so will have to be made available.

What we are witnessing at this time of transition are the beginnings of a new awakening on the part of *some* of the leaders in our society to the new realities. High technology companies are taking the lead, taking new initiatives to found partnerships with academia. A few state governments are waking up to their new responsibilities. Many college and university officials are willing and even anxious to move ahead. What is clearly lacking on the scene is any sense of direction or even understanding on the part of the federal government.

In the next chapter, we ask what the federal government might do to build upon and sustain the initiatives that are clearly evident from other sectors of society. We make no appeal for a national plan such as those that exist in Japan and France. But we do argue that only the federal government can bring the necessary leadership and sustaining power to a situation that requires some form of national policy.

Chapter 9

IN SEARCH OF NATIONAL POLICY

THERE IS no easy formula for pulling all the pieces of a national policy together. No magic wand. Establishing a new economic strategy seems doubly difficult in the present domestic budget-cutting, defense-minded atmosphere of Washington. But an opportunity is taking shape. Industry is moving toward a closer relationship with the educational institutions upon whose resources it relies. In turn, academia is proving more responsive to the evolving needs of an industry in rapid transition. Two legs of a stool are thus being put into place. But to provide firmer footing, a long-term sense of direction and a dependable funding mechanism must be added.

The missing third leg is federal support. The high tech industry can initiate new partnerships with education, but only a concerted national effort can sustain them. Precedents in American history can help define a role for federal policymakers. One of them occurred more than a century ago when Abraham Lincoln signed into law the historic Morrill Act, the land-grant college legislation in 1862, which set the stage for a revolution in American agriculture, engineering, and U.S. education. More importantly, it set the nation on a new economic course and established a precedent for the present. What the United States needs now is to marshal the political will to again set a coherent national policy for the coming decades.

A High Technology Morrill Act

More than a century ago a man named Justin Morrill, then a congressman from Vermont, sponsored a federal act founding colleges to "teach such branches of learning as are related to agriculture and the mechanic arts." This led to the establishment of the agricultural extension programs and to the birth of modern farming in the United States. If American agriculture plays such an important role in the world economy today – it leads the world in agricultural exports while Japan, for example, imports 50 percent of its food needs and Western Europe imports 25 percent – it is not just because its land is rich and fertile or its farmers hardworking. It is in part because the Morrill Act fused the interests of government, education, and the farming community into a national policy. This educational act endured and remains as a compelling model for the future of American high technology.

THE LEGISLATION (1862)

The Land-Grant Act of July 2, 1862 (First Morrill Act) signed into law by President Abraham Lincoln stated that the act would provide for:

the endowment, support, and maintenance of at least one college where the leading object shall be, without excluding other scientific and classical studies and including military tactics, to teach such branches of learning as are related to agriculture and the mechanic arts, in such a manner as the legislature of the States may respectively prescribe, in order to promote the liberal and practical education of the industrial classes in the several pursuits and professions in life.

Proceeds from the sale of lands "shall be invested in . . . safe stocks, yielding not less than five percentum upon the par value of said stocks; and the moneys so invested shall constitute a perpetual fund . . . the interest on which shall be inviolably appropriated" for the purposes mentioned above.

(Sec. 4 [as amended April 13, 1926] 44 Stat. L. 247)

As part of the Morrill Act, the federal government donated 17,430,000 acres of land to help subsidize the founding of a system of colleges. These were to make judicious use of the revenues of this land to support their educational programs. A second Morrill Act in 1890 added the germinal concept of federal incentive grants to states. In his highly respected history, *Education in a Free Society*, S. Alexander Rippa notes that "the Morrill Act not only laid the foundation for a new type of curriculum at government expense but also provided powerful incentives for greatly expanded state programs of higher education."[1]

To this day, the agricultural land-grant college system remains a cornerstone to the continued efficiency and proficiency of the agricultural economy of the United States. It came into being at a time when industrial methods were just beginning to revolutionize production on the farm. As late as 1860, agriculture remained primitive at best. This would soon change. Between 1855 and 1895 the hourly labor required to produce one bushel of corn declined from four hours and thirty-four minutes to forty-one minutes, while the time required to produce a bushel of wheat declined between 1830 and 1894 from three hours and ten minutes to only ten minutes.[2] The new land-grant colleges would sustain this momentous productivity in American agriculture.

SOME UNEXPECTED ASPECTS OF THE LAND-GRANT ACT

One of the earliest recipients of support under the act was the Massachusetts Institute of Technology. In 1863, "the legislature of Massachusetts passed three acts — one accepting the land-grant, one incorporating the agricultural college [the University of Massachusetts at Amherst], and one dividing the land-grant proceeds between the agricultural college and the Institute of Technology in a two-to-one proportion."[5]

Chartered as an educational institution in 1861, MIT waited out the Civil War years before formally opening for classes in 1865. Since then, it has continued to be a recipient of land-grant funds. The 1981 federal land-grant payment to MIT totaled $23,000.

When the land-grant act was passed in 1862 there were five engineering schools in the United States. Twenty years later there were eighty-five, half of them engineering departments at land-grant institutions.

"Among the thousands of land-grant research discoveries have been . . . fundamental work in *transistors* and on the first *television tube* at Purdue University; the invention of the *cyclotron* at the University of California." Edward D. Eddy, Jr.

The largest university system in the United States, the University of California, is a land-grant institution.

One hundred years after the act was passed, one out of four higher education students was enrolled in a land-grant college or university.

Part of the land-grant act called for training in "military tactics." This requirement gave birth to the Reserve Officers Training Corps — better known as ROTC.

Perhaps the most impressive legacy born of the Morrill Act was the understanding that education — open to all and focused on learning applied to real economic needs — could not be divorced from economic growth and national strategy. Education became a matter of national priority, one to which the federal government would devote its resources. It was a revolutionary milestone in turning the United States away from a legacy of British and German educational philosophies founded upon mastery of the classics. The French chronicler of American thought, Alexis de Tocqueville, foresaw this difference between U.S. and European education. He said: "It is evident that, in democratic communities the interest of individuals, as well as the security of the commonwealth, demands that the education of the greater number should be scientific, commercial, and industrial, rather than literary."[3]

A Four-Sided Challenge: Creating a national economic policy presents at least four challenges. There is a need for sustained financial support, life-long education, high school incentives, and a global view of technology.

The federal government needs to provide dependable, *sustained financial support* to the American system of education. The high social value of investment in education is apparent to all. Yet the fraction of this value that individuals are able to capture as their own return is too small to enable the entire support for education to come from indvidual decisions alone. Some significant share of the funding must come from the aggregate national income or wealth.

A "High Technology Morrill Act" would fuse a new partnership among federal, state, and local interests. The keystone of such a legislative act would be to respond to *matching grant* initiatives from state governments and from high technology industries. This would not only add to the leverage of federal educational appropriations, but more importantly, it would ensure that federal allocations respond to the lead of industries and states in selecting those technological fields where the educational investment could be maximized. Thus, for the electronics industry in Michigan, the bias might be toward robotics engineers; in California it might support semiconductor electronics research; and in Massachusetts it might focus on computer software engineering or semiconductor manufacturing processes.

Encouraging a larger flow of money from industry to universities would help to assure closer connections and cooperation at a time when the distinctions between theory and practice need to grow smaller. It would encourage universities to teach things of practical value to students. Over the very long term, the participation of industry would allow a dynamic process to occur: the sources of funding would shift if the high tech industry's ability to sustain the matching process matures and is superceded by a new generation of leading edge industries.

If, for illustrative purposes, the estimated annual matching-fund limits were arbitrarily set at $1 billion to maintain technological state-of-the-art education facilities in the nation's 300 or so engineering departments, the matching challenge might be based on a 5-3-2 formula: 5 federal dollars, for 3 state dollars, for 2 industry dollars. On an aggregate basis:

Federal commitment year 1:	$ 500,000,000
State commitment year 1:	$ 300,000,000
Industry commitment year 1:	$ 200,000,000
	$1,000,000,000

Unlike the 1862 Morrill Act, which sought to establish new educational institutions throughout the nation, the High Technology Morrill Act would be focused on strengthening and sustaining existing institutions. In addition, the new high technology act would be designed to function from the bottom up. An initiative by a corporate donor would then be matched by state and federal funds in a 5-3-2 formula; the donor might be matched only on the amount that exceeds amounts donated in prior years. This would intentionally bias the process toward high growth, sunrise industries that can afford incremental donations by virtue of their increased revenues and profits. In this manner, the new industries such as electronics, biogenetics, and other high tech ventures would automatically induce leverage by their own donations.

One cannot overlook another precedent of the Morrill Act. Through amendments to the original act, the vision of education as a lifelong endeavor came into being. The experimental agricultural research stations that were established at selected land-grant institutions undertook not only to push research to new frontiers but to educate the farmer in an ongoing, lifelong manner. It is ironic that more than 100 years later, the concept of continuing adult education or lifelong cooperative learning is considered new ground for educational institutions. The precedent is there and has to this day maintained a massive effort to teach the adult farmer new techniques and new crop potential. Through an extension of the Morrill Act, this experience can be adapted to the needs of American high technology society.

National policy must reverse the dismal conditions now visible in many, some say most, of the nation's 17,000 elementary and high school education systems. This must be done with vision and determination. National policy must take a long-term view of the responsibility to educate its young for a more complex future.

Serious shortcomings of high school education in math and the sciences should receive special consideration in the formulation of a new High Technology Morrill Act. For every funded program on the campus of a university, college, or community college, a fraction of the allocated funds might be assigned to one or more high schools, or to an elementary school program. Equipment grants for high schools, exchange programs between college and high school students, and special training services for high school and elementary school teachers are but a few of the links that might be incorporated into a new Morrill Act. Again, whether or not this is the appropriate mechanism, the point is that national policy should couple the needs of elementary and secondary education with those of higher and continuing education.

Finally, the question of international perspective deserves special consideration. Global markets are bringing new relationships into focus as high technology industries mature. The rapid pace of technological change is causing an ever wider breach between the interests of the United States and those of many of many of its trading partners, especially in the Third World. We may chose to ignore that our world is increasingly interdependent; or we may seek to find ways to understand and promote international cooperation.

High technology and its educational requirements represent both opportunities and problems for many Third World countries. Proper use of information technologies in ways that respect local culture has a potentially positive role to play in furthering economic and social development, especially in the newly industrializing countries like Brazil or Singapore. On the other hand, the requirements both for financial capital and scientific and managerially trained talent is presently beyond the reach of the poorest of the developing countries like Bangladesh.

It is in the long-run interest of the United States to nurture healthy partnerships with developing countries as well as with other advanced economies. Part of a new Morrill Act could be aimed at bridging the gaps between developed and less-developed countries and to use the information technologies to narrow rather than widen the economic disparities. For example, sponsoring a forum for international leadership and educational exchanges could encourage a two-way transfer of technology and ideas. This would be one way to help maintain fruitful long-term economic and political relations as we enter an era of global markets and a world economy.

The Missing Piece: National Strategy

A national strategy must begin with an unabashed and strong commitment by the president of the United States. Not only must he articulate a vision of the future but he must craft long-term goals that account for both the knowledge-intensive nature of the economy and the international pressures bearing on it. Such a policy must communicate to the American people the fundamental role education plays in strengthening our economy and culture.

The president must also strengthen his science and technology advisory structure. One way to start this might be to set up a Presidential Commission on Technology and Productivity. This group might consider proposing to establish a standing committee or advisory group in the White House to deal with these issues. By chairing the activities of a science and technology advisory group, the president would not only transfer his authority to it but would also participate actively in setting its agenda. This would announce emphatically to the nation that the president is placing the highest domestic priority on science and technology.

In addition, such an advisory group should, in the words of a National Governors Association,

> serve as the deliberative body responsible for defining federal goals and objectives relating to science and technology policy, [to ensure] that those processes, public and private, essential to realizing the benefits of science are [achieved]. The group would define goals and objectives germane to U.S. society, but the analogy is with the deliberate decision by Japan to take "extraordinary measures for the promotion of the electronics industry."[6]

Such a commitment would require that the president make a clear statement of economic goals and their relationship to science and technology. Thus one might expect the president to speak bullishly of the future of robotics technology in America, or of a new era of semiconductor technology, or of an American determination to lead the world in biogenetics, all within the context of maintaining a dominant role for the United States in a global economy.

Another fundamental aspect to national policy rests with Congress. A legislative initiative redefining the Morrill Act of 1862 into a High Technology Morrill Act would restore vitality to the nation's engineering and science departments and laboratories without sacrificing the original intention of the act to educate a well-rounded student with a strong foundation in the humanities. Such a measure would put education back into the central arena of public policy to which it was raised for a few short years following the Sputnik scare during the late 1950s and should bring into being a long-term funding mechanism.

The will is there. In industry. In academia. In an increasing number of state governments. The missing piece in a game of global stakes is a coherent, visionary, and articulate national policy for a knowledge-intensive era. What is needed is a clear commitment to develop the human resources upon which economic growth and development are dependent. A responsive national leadership must look to initiatives emerging from a decentralized, market dependent, free economy. And when federal funding is committed, it must complement the initiatives of industry, academia, and state government. A true national strategy is one in which all the key players work in concert.

These thoughts are only a beginning. Our goal was not to suggest a comprehensive solution to the complex challenge we face in high technology, but rather to contribute to a deeper understanding of the issues involved and to encourage a more active dialogue on this important subject.

The forces for change must be initiated by those who are most knowledgeable and directly concerned. This starts with the high technology industry itself as well as with those universities that have a vested interest in education in science and technology.

But any action must also include state governments which must be concerned about future employment of their citizens and about competing for their share of the knowledge based industries. The present reality is that, in the aggregate, state governments have the greatest control over and potential impact on higher education. They can significantly influence investments in university research. But with very few exceptions, they have so far failed to exercise this power in a way that directly influences targeted economic development objectives. So not only does our federal government need to sharpen its strategic economic development goals, but state governments need to greatly increase their strategic planning skills and to learn how to work more harmoniously with industry and universities in establishing long term development goals.

It is encouraging to see the bottoms-up approach for change that is welling up from the constituencies most directly affected and concerned. But it is difficult to envision how these efforts can be successful without a coherent national policy and leadership to bind these uncoordinated efforts together and to provide a proper allocation of federal funds to strengthen our system of education.

Chapter 10

THE FUTURE OF HIGH TECHNOLOGY

THE SUBJECT of this book is complex. In the first nine chapters we have woven a broad canvas of the issues and some of the potential solutions.

Because the subject of high technology, economic growth, education, and defense policies affect all levels of our national life, we have invited a number of individuals to share their views from their respective institutional vantage points. These views include those of The Honorable James B. Hunt, Jr., governor of North Carolina; John Young, president of Hewlett-Packard Company; Dr. Paul Gray, president of MIT; Erich Bloch, chairman of the Semiconductor Research Cooperative and an IBM executive; Kenneth G. Ryder, president of Northeastern University; and Professor Robert M. Hexter, director of the Microelectronics and Information Sciences Center at the University of Minnesota.

A closing contribution from France by Jean Saint-Geours, a special advisor to the prime minister, offers a further valuable insight into the future impact of electronics and knowledge economies on society.

ACADEMIA, INDUSTRY, AND GOVERNMENT: THE ORGANIZATIONAL FRONTIER OF SCIENCE TODAY

by

The Honorable James B. Hunt, Jr.
Governor of the State of North Carolina

As a governor, I feel very keenly the need to stimulate vigorous dialogue between scientists and engineers on the one hand and political and industrial leaders on the other. Without that dialogue, I believe that we simply cannot achieve the goals we set for ourselves in North Carolina and in this nation.

Nearly twenty years ago James R. Killian, then chairman of the MIT corporation, addressed a conference of state governors held in Miami. He began as follows:

> Much has been spoken and written in the last decade and a half about science and government, and the discourse has dealt almost exclusively with the *federal* government. . . . At MIT, I occasionally give a graduate seminar on science and public policy, and I must admit that state governments are hardly mentioned, an omission marking similar seminars in other universities and most of the books published on the subject.

Dr. Killian goes on to say that the governors' conference he was addressing was a "start in redressing this neglect" of science in state governments.

Unfortunately, state governors have not made the progress expected twenty years ago. But neither has the federal government nor the private sector. The glaring fact is that, in the United States today, we are not realizing the full creative potential of science and technology. The creative potential of science, that is, the potential benefit to society, is achieved through technological innovation. I believe, however, that a serious crisis is now emerging in the United States because we have not mastered the processes of innovation, as have Japan and a few other nations.

Technological innovation consists of two interrelated parts: technical and organizational innovation. Technical innovation is developed through the use of physics, chemistry, biology, and other natural sciences, plus mathematics and engineering. Such knowledge makes it possible to produce new or better physical or biological products, or produce them more efficiently.

Organizational innovation is accomplished through the use of economics, political science, sociology, and other social and behavioral sciences. The objective is to change the organization and operation of units of society. Organizational innovation can occur on a large scale, such as at the national level – or it can be on a small scale, such as in an office or industrial firm.

Professor Don Price of Harvard University, in his survey of the World War II period, says: "The most significant discovery or development for science and technology to come from the war effort was not the technical secrets that were involved in radar or the atomic bomb. It was the administrative system and set of operating policies that produced such technological feats."[1]

In other words, the organizational innovations developed on a national scale during and following World War II were as important, if not more important, than the technological innovations of radar and atomic weaponry. But each is dependent upon the other. We cannot reap the benefits of science without both.

I suggest also that proper study of technological innovation, as I have defined it, is as challenging as genetic engineering or cellular biology. Forms of scientific and engineering knowledge, like genetic codes, influence the behavior of components of society, and, like the cellular structures of higher organisms, the interactive behavior of these components give form and character to society.

The importance of technological innovation arises from the fact that no modern society can function as such today without making effective use of scientific knowledge. Our military strength, as Price has noted, depends upon it. But of greater significance to the emerging crisis today is the dependence of economic productivity upon our ability to innovate both technically and organizationally. The problem, as I see it, is that we still have essentially the same organizational structure for science and technology that was designed in World War II and the Korean War.

Even the establishment of the National Science Foundation in the early 1950s was influenced by the war experience. And the burst of federal support for science that followed Sputnik in the late 1950s was inspired more by national security concerns than by the altruistic potential of science for all humanity.

Federal expenditures now dominate all research and development in the United States — $42 billion out of a total of around $70 billion. Within the federal outlays, defense-related expenditures are by far the most dominant category — $24 billion out of $42 billion.[2] Thus, in several respects, it may be said that science and technology have ridden the coattails of defense and space since World War II.

The unique aspect of military and space research is that provision is made for technological innovation by actually producing and using weapons systems, rockets, tanks, ships, and guns. The Department of Defense and the National Aeronautics and Space Agency organize to use the results of research in fulfilling these federal objectives. For fundamental nonmilitary research, little provision is made for effective utilization of results — the critical fault of our present system.

The Emerging Innovations Crisis:

A great deal has been accomplished over the last thirty to forty years by our prevailing structure of science and technology. We have won wars. We have more Nobel Laureates than any other nation. We have shared our technology through foreign assistance. And for a time we were the marvel of the world with respect to scientific achievement and technological advancement. Now, however, certain deficiencies are becoming apparent:

Economic productivity. Output per man-hour in the United States has leveled off in recent years; in some years, it has declined. A recession now prevails, inflation continues, and unemployment is rising. Moreover, we are now deliberately seeking to control inflation by slowing down economic growth rather than by measures designed to increase productivity.

Fundamental research accomplishments no longer percolate through our economy with sufficient rapidity and effectiveness to substantially increase productivity. Japan, for example, drawing heavily upon our own research, is exceeding our ability to transform such fundamental knowledge into useful products with worldwide market potential.

Education. U.S. education, especially at the elementary and secondary levels, is significantly less rigorous than that of several other nations at the present time. Japan, for example, has the highest literacy rate in the world. In the Soviet Union, five million students graduate from secondary institutes each year, having had two years of rigorous instruction in calculus. By comparison, of all our high school graduates each year, few more than 100,000 complete one year of calculus.[3]

Our workforce as a whole is not keeping pace with the needs of a highly technical society. Critical shortages exist, not only of well-trained engineers but of engineering faculty. Shortages are emerging in relation to other areas as well, such as biotechnology. Financial support for universities, the status of scientific equipment and research facilities, and the opportunities for young faculty, all these and other measures of the vitality of our academic institutions are discouraging.[4]

Environmental Management. A recent New York Times/CBS News Poll reports that "a large majority of the American public supports continued strong protection of the environment even if it requires economic sacrifice." Clearly, we have the scientific capability to manage land, water, and air resources properly and to minimize the dangers associated with toxic, hazardous, and low-level radioactive waste. But we have not yet devised the organizational means to generate and use such knowledge adequately and effectively.

Views of the American Public

According to the pollster Louis Harris,[5] Americans want inflation curbed, government expenditures cut, economic productivity increased, our technological capability greatly enhanced, and America's standing in the world restored. They feel that government performance has been costly and often ineffective, that industry has concentrated upon minor short-run gains at the expense of major long-run breakthroughs, that the scientific capacity of our universities is ten years out of date, and that things should begin to turn around by the end of 1982 if the policies of the Reagan Administration and American business are going to prove effective.

But Harris concludes that reliance by the Reagan Administration upon free market forces will not right the economy, that the private sector may not be up to the task, and that American patience will be stretched to the breaking point within a year. Resolution of the crisis of confidence will not be through reversion to New Deal economics. Instead, a redefinition of the role of government will be essential.

Innovation and the Role of Government

The emerging crisis presents a challenge to our ability to innovate comparable to what we faced in World War II. The overriding question now, as then, is how to organize our remarkable scientific and engineering capability. Now it should be for peace; then it was for war.

The situation both then and now demonstrates the central importance of science and technology to this or any society. It follows, therefore, that those who devise the organizational means to resolve this emerging crisis — who successfully redefine the role of government in relation to academia, industry, and the general public — will also reap the rewards of economic and political support inherent in the fundamental desires of Americans throughout this country.

Leadership and Style

In redefining the role of government in relation to research institutions, industry, and the general public, I contend that *the center of gravity for technological innovation must shift from the federal government to state governments.* This will require greater differentiation between the role of the federal government and the roles of state and local governments. And the style of government at each level will need to be catalytic, fostering the creative spirit of technological innovation throughout society, and not just down from the top.

The Federal Role

Before outlining the state role, I should note that the federal government must play an important supportive role. Elements of support should include: (1) maintaining a favorable economic environment by controlling inflation and ensuring adequate flows of capital at reasonable interest rates; (2) providing tax and other incentives designed to encourage investment in research and development; (3) strongly supporting basic research relevant to all states; and (4) assisting groups of states in uniting to pursue common research and development interests.

The point to be stressed is this: Measures such as these comprise the role of the federal government in fostering a broad-based program of technological innovation across this country. The federal role is not to "administer" innovation through the same style of leadership that is followed in devising a new weapons system or in putting a man on the moon.

The Role of State and Local Governments

Much is spoken these days of a partnership of government, academia, and industry. I believe that such a partnership is essential, and that the best way to organize it is through *state* government leadership.

We should remember that, of the 184 research universities of this nation, 119 are public institutions, most of which are supported by state governments. Many of the remaining private institutions receive some form of state support. Virtually all university research is conducted by these 184 institutions, and they exist in every state. State governments are committed to higher education but, for the most part, they have not learned how to foster and utilize in a catalytic fashion the tremendous research capacity of our academic institutions. This must be remedied. Academia is where we train our scientists, engineers, and technicians. It is where scientific exploration takes place that makes technological innovation possible.

We should remember also that elementary and secondary educational systems are the responsibility of state and local governments. Science and mathematics instruction at these levels of education needs extensive and striking improvement. Regardless of action by the federal government, significant improvements will be achieved only if state and local governments take the lead in doing so.

We should remember too that industrial firms, farms, banks, wholesale and retail outlets, transportation systems, and all other forms of economic activity exist and function within state boundaries and relate closely to state and local governments. These units of government are the prime points of contact with respect to locational issues, labor relations, environmental management, provision of capital (both human and material), living conditions for employees, and other facets of economic activity that entail industry-government interaction.

We should remember, in addition, that *people* live within state and local government jurisdictions. I include people as an absolutely essential partner in technological innovation because the entire system must work in their interest. We cannot regard people as simply another abstract factor of production called labor, a factor that is used and discarded at will, and for which the system has no overall responsibility. Our children, our youth, our workers of every type in their most productive years, and our senior citizens are all indispensable participants. People can relate to state and local governments more easily than to a distant federal agency. Therefore, if state and local governments play a more significant role in technological change, that change is likely to be more responsive to the desires of people, particularly if the partnership reaches out to include them.

Examples of State Leadership

To illustrate how a state government can forge partnerships, let me share some of the experiences of North Carolina and a few other states.

First, certain state-level organizational arrangements are essential. The North Carolina Board of Science and Technology is the unit that maps much of the strategy by which we are proceeding, building upon the work of our universities and the influence of our Research Triangle Park. As governor, I chair this fifteen-member board; the remaining members are scientists from our public and private research institutions and officials from state and local government.

I also have other groups that advise me and help develop the essential working relationships. One, for example, is a council of business leaders from North Carolina. Our North Carolina Department of Commerce works closely with this group and other institutions, firms, and individuals within and outside North Carolina. As a consequence, new industrial investment in North Carolina has averaged approximately $2 billion per year for the past five years. Our unemployment rate is running about 2 percent below the national rate as the current recession unfolds.

Some twenty states now have organizational arrangements resembling those of North Carolina. Another ten are beginning to initiate such arrangements. The remaining twenty appear to be less active at present but are exploring possibilities.[6]

Next, in North Carolina we are investing in people, particularly young people. In our elementary and secondary schools, we have introduced competency testing, raised teacher pay and improved their training, reduced class size, and taken other measures to improve education in general. Significant improvements in national test scores are one indication that these changes are having an effect.

In addition, we have established the North Carolina School of Science and Mathematics. This is a residential high school for students with very high aptitudes for science and mathematics. The purpose of the school is twofold: (1) to train and inspire those students-in-residence to become future leaders of science, and (2) through outreach programs, to help upgrade science and mathematics instruction in all elementary and secondary schools of the state.

Now in its second year, the school has 300 students. No more than 900 will be in residence when it reaches full capacity. About 15 percent will be from out of state. I should add that about an equal number of males and females are enrolled, that the distribution of students by race is proportional to that in the state, and that students are drawn from small rural schools as well as from large city schools.

It is too early to gauge effects, but in its first year with 150 students enrolled, this school had the second largest number of 1982 National Merit Scholarship semifinalists of any school in the nation.

My last example consists of our Microelectronics Center and our Biotechnology Center. The Microelectronics Center is designed to enable six of our leading research institutions in North Carolina to have access to very sophisticated microelectronics research equipment on a sustained basis. This will make it possible for these institutions to remain on the frontier of research and education in this field. Such equipment is extremely expensive and becomes obsolete quickly. We established the center in order that these institutions may share this equipment. Otherwise we could not afford to equip each at this level of sophistication.

Further growth of microelectronics firms, and of firms that use microelectronics products, is expected in North Carolina as a consequence of the availability of highly trained engineers and other professionals, plus the advantage of being near vigorous research activity.

Our North Carolina Biotechnology Center is beginning on a relatively small scale, but represents a long-run commitment to this field. We recognize that many financial, patent, and other issues must be explored carefully as we develop closer working relations between industry and our research institutions and government.

Our intent is for North Carolina to become a national leader in biotechnology research and development. Our key research institutions, working closely with the center and with our Department of Commerce, are planning major developments in research programs, faculty, and facilities in order to achieve this goal.

Other states are beginning to take action in relation to fields of exploration such as microelectronics and biotechnology. California has initiated a program of financial support to research and development in microelectronics. The states of Minnesota and Michigan are also taking significant action. Florida is providing significant funding for biotechnology. There are many more examples to show that the nature and extent of state government activity is increasing steadily.

You may ask whether state governments are able and willing to provide financial support to such activity. Let me assert emphatically that I believe that any legislature will provide strong support for research and development if you demonstrate that the state will indeed benefit from the research. I do not subscribe to the belief that total support for research and development is fixed, and that if we increase expenditures for technology we take it away from basic research.

I make this assertion in spite of the fact that total appropriations by all fifty states for directed research is now less than $500 million per year. This is little more than 1 percent of total current annual research and development expenditures by the federal government.

The North Carolina Legislature, however, recently appropriated $24.4 million for our Microelectronics Center. Why? Because the members of the legislature saw the connection between the center and better jobs for our citizens.

Our legislature is also providing nearly $20 million per year for agricultural research and development, four times the federal expenditure for agricultural research in our state. Why? Because members of our legislature know that we cannot increase agricultural production without such research. The dollars we spend on agricultural research result in *increased* production on our farms. Translated into return on investment in research, each such dollar earns between 35 and 70 cents per year in increased productivity, a very good rate of return on investment indeed.[7] Farmers know this and farmers vote. Legislators know that farmers vote.

A broad-based program of technological innovation will be, among other things, that process by which citizens see and feel the benefits of research. If states deliberately ensure that the results of research are used for the benefit of citizens, legislators will provide the support. Citizens will insist that they do so. Congress will help.

Guiding Principles

Technological innovation should be construed as more than an end in itself. The larger purpose of innovation may be expressed as "meeting the needs and desires of people." Determination of needs and desires is a function of the values and beliefs of people and of political and economic processes. The emerging crisis I have described is a reflection of such concerns and desires. Government – particularly state government in partnership with academia, industry, and people – has a clear responsibility in resolving this crisis.

Any partnership, however, runs the danger of becoming a collusive arrangement whereby one or more partners take advatange of other members, or whereby all partners conspire to take unfair advantage of those outside the partnership. Therefore, principles of integrity and purpose must be agreed upon explicitly. Provision for debate, negotiation, and periodic review are essential. To forge the partnership required, and to establish durable operating principles, each state will likely find it necessary to reach beyond the domain of science and involve those versed in other relevant fields of knowledge and experience.

Let me put this reasoning in philosophic terms: When we forge and implement an important policy decision affecting much of society, we integrate forms of knowledge, factual evidence, values, and beliefs. Good or bad, a policy decision is what it is. But I operate under the assumption that our policy decisions are better decisions when we make full and effective use of scientific knowledge. This is why I feel so strongly about the importance of our scientific and engineering community. It is also evident, however, that integrity, purpose, ethical and aesthetic values, and other considerations are essential in policy formation.

Much has been said and written about what the Reagan Administration calls the "new federalism." Like it or not, the relationship between the federal government and states is undergoing a radical shift, a shift that has far-reaching implications for the advancement of science. This is the organizational frontier of science today. In light of these new realities, it is more important than ever that state governments take a leadership role.

AN AGENDA FOR THE ELECTRONICS INDUSTRY

by
John A. Young
President
Hewlett-Packard Company

As many American industries struggle with the problems of diminished growth and lagging productivity, the electronics industry continues to be a strong performer. Worldwide sales of U.S. electronics companies increased at an average annual rate of 17 percent in the last decade and, in 1980, amounted to more than $104 billion. Of those sales, more than a third were in the electronic equipment and systems category. Next, in order of volume, were communications equipment and systems, electronic components, and consumer electronics.

It is estimated that, by 1990, the electronics industry on a global basis will have jumped from the tenth largest to the fourth largest industry in the world. By the year 2,000, it will be second only to energy.

Fueling such rapid growth is the continuous flow of new and innovative products to the electronics marketplace. At Hewlett-Packard, for example, about three fourths of our total business in 1981 was derived from products introduced since 1977. Not only new products, but entirely new product areas are being developed. Many of the fastest-growing areas were practically unknown ten years ago: personal computers, advanced handheld calculators, new forms of telecommunications devices, automated office equipment, and others.

Semiconductor technology provided the base for most of these products. For the past twenty-five years, the real cost of an electronic function has been halved every two and a half years — a notable achievement in today's world of inflating costs.

The importance of such innovations to the general well-being of the

United States cannot be grasped simply through the growth figures of the electronics industry itself. Even more significant are the increases in industrial productivity being generated by computers and other electronic products. Individuals can now do their jobs more efficiently and more economically than ever before. Since U.S. electronic products are in strong demand overseas, they also contribute to a favorable balance of trade for the nation. Furthermore, they play a key role in the country's national defense, where development and control of strategic technology increasingly determines the international power alignment.

Underlying the success of the domestic electronics industry has been an absolute faith in American ingenuity and entrepreneurial spirit. For decades, the United States has stood unchallenged as the world's fountainhead of technical innovation. Knowing this has given Americans a sense of pride and confidence in the systems that have created the nation's technological leadership.

Today, this once-secure position at the top of the heap is under assault. From Japan has come a steady stream of technologically advanced, high-quality, low-cost products that seriously challenge American goods. In Europe, government investment in the electronics industry and a determination to influence local markets have reduced the competitiveness of the marketplace and made it more difficult for American suppliers. A sense of purpose – one that clearly ties success in the electronics marketplace with the national interest – is rapidly emerging in many countries.

It is time to take a long, hard look at the challenge from abroad. How serious is it? Are the mechanisms in place in the United States to maintain technological leadership? Or will the American electronics industry – like many other industries – fail to react in time?

Careful study of the situation leads me to two principal conclusions: (1) foreign competition in its various forms is indeed serious and meeting it will require our best strategic response, and (2) such a response will require better management, improved research and development effectiveness, and new expectations for quality. A necessity in all these areas is the ready availability of well-trained engineers and computer scientists. Achieving this goal will require well-directed, cooperative programs among industry, government, and educational institutions.

Ample evidence exists to support the first conclusion. One clear example is the success Japanese semiconductor companies have had in capturing market share in computer memory components. By investing skillfully and taking advantage of a temporary plant-capacity shortage among U.S. component manufacturers, Japanese suppliers were able to enter the market for a particular memory product in 1976 and claim 40 percent of it by 1980. With that foothold, these same suppliers have captured an estimated 70 percent of the market for the current-generation memory component.

Instant market dominance such as this doesn't just happen. Extensive testing of Japanese memory parts in 1979 demonstrated that they were clearly less likely to fail under operating conditions than parts from U.S. suppliers. While American companies have made great strides in closing the quality gap, a significant incremental engineering investment was required.

Japanese government support for technological achievement is unwavering. In a report dated March 1980, Japan's Ministry of International Trade set forth a national goal for the 1980s "to develop innovative and original technology." Pursuing that goal, the Japanese have formulated ambitious plans to capture a major share of the worldwide computer market by 1990 with an advanced fifth generation computer design now under development.

Other nations are not standing still. Aware of the economic and social benefits of an indigenous electronics industry, many European countries are supporting their firms not only through direct government aid, but also through tariffs and other protectionist policies. These barriers limit access of non-national firms to local markets and threaten the whole concept of free international trade.

If foreign competition is indeed the threat suggested, what then is this nation's best strategic response? Erecting our own trade barriers? Subsidizing industry? Underwriting new ventures?

None of these gets to the root of the problem. If we get down to fundamentals, a basic element that has clearly sustained the U.S. technological leadership over the years is our educational system — the source of the electronics industry's greatest strength: its people. Looking at this system today, we find several problem areas that, if given proper attention, could go a long way toward preserving our leadership.

One shortcoming of our university system today is its inability to turn out sufficient numbers of engineering graduates. The American Electronics Association suggests that the U.S. educational institutions will have to triple their output of electrical engineers and computer scientists within the next five years to meet the projected needs of the electronics industry alone. By contrast, Japan – our strongest competitor in electronics – turns out about two and a half times as many engineers on a per capita basis as we do.

This shortfall of engineers in the United States is not due to a lack of students who want to study engineering. The problem is that there is not sufficient room for them. Many engineering departments at public universities are turning away more students than they can accept.

Those students who are enrolled in engineering are quickly faced with a second problem: antiquated equipment. In a profession that thrives on state-of-the-art developments, young people frequently are being taught in outdated facilities with equipment that is older than they are. Reduced government funding for universities and the high cost of modern equipment makes it increasingly difficult for even the leading universities to keep up.

The third and perhaps most pressing problem is a severe shortage of engineering teachers. Of the 16,200 positions available on U.S. engineering faculties in the fall of 1980, 10 percent were unfilled, according to the American Council on Education. The shortages in such rapidly growing areas as computer science, digital systems, and solid-state electronics have reached 50 percent in some schools.

Low, uncompetitive faculty salaries are a principal cause of the teacher shortage. Fewer engineering graduates are choosing to enter doctoral programs, and many of those who have earned doctoral degrees are nonetheless opting for positions in industry, where holders of bachelor's degrees can begin in the $24,000 to $26,000 range (1981 data) versus the $21,000 to $23,000 salaries generally offered to assistant professors. Moreover, many veteran engineering teachers can realize immediate increases in pay by jumping to industry employment. Obsolete research equipment at many universities provides another disincentive to teaching. These conditions help explain why some estimates show 3 percent of existing engineering faculty migrating to industry jobs each year. The resultant decreases in faculty-student ratios, of course, diminish further the quality and rewards of the educational process.

What actions are needed to address these three problem areas in our engineering schools? Obviously, a broad effort must be mounted, one that requires the cooperation of leaders in industry and government as well as education.

One of the best examples of such cooperation is the Stanford University Center for Integrated Systems already introduced in an earlier chapter of this book. The center is expected to produce 100 master's and 30 Ph.D. graduates each year.

Legislative action at both the state and federal levels can encourage such philanthropy among private companies, acting alone or in combination with each other. Especially encouraging to private industry are the incentives that liberalize write-offs for equipment donations and offer credit for money spent on research and development, whether companies spend it on their operations or grant it to universities. The American Electronics Association has urged its members to contribute to education 2 percent of what they spend on research and development each year.

Along with the problem of outdated equipment, the engineering faculty shortage is also receiving industry attention. Several short-term, practical measures have expanded faculty resources. The use of privately owned closed-circuit television to transmit university lectures beyond crowded classrooms is one example. This permits engineers in private industry to continue their education by taking classes at or near their workplace. Companies are also making experienced engineers available to universities as teachers so that course offerings can be increased without the hiring of additional faculty. Professors are being invited to spend their summers or sabbaticals working in industrial facilities, or to form consulting relationships with industry as a way of adding to their income and linking industrial and academic research.

Yet the funds needed for better equipment and higher faculty salaries cannot come from industry alone. Education is an economic and social asset that government must always support. Scarcity of money for graduate fellowships and faculty research grants will continue to force more bright people to abandon universities, unless something is done.

It seems entirely consistent with the U.S. national interest that the government not only preserve, but strategically increase its funding of engineering education. In the debate over tax breaks, it is important that legislators recognize the relatively high cost of funding effective engineering education compared with other, less capital-intensive academic endeavors. In addition, it is important that they recognize the vital, long-term benefits, particularly the job-creation capabilities, to be derived from electronics and other high technology industries.

We are on the threshold of one of the most exciting periods in the history of high technology. There is a strong challenge facing us from other parts of the world. Whether or not we can retain our leadership will depend on how well we meet this challenge. The potential rewards can be measured in jobs, in a more positive balance of trade, in an abundance of products and services that enhance the quality of life, and in a stronger posture in national security – in short, in the well-being of every American. If industry, educational institutions, and the government can combine to produce the people and the environment needed for the task, our success will be assured.

ENGINEERING EDUCATION: THE PROBLEM

by

Dr. Paul E. Gray
President
Massachusetts Institute of Technology

In the not-too-recent past, the high quality of engineering education in this country could almost be taken for granted: engineering schools were growing with faculty, with students, and with new intellectual horizons. Today, the national and regional system for educating engineers is at saturation. Its expansion is limited by several fundamental factors, of which the most important are constraints on the ability to renew and enlarge engineering faculties. The most probable outcome of this growing problem is a significant decline in the quality, if not the quantity, of engineering graduates in the years ahead.

A historical snapshot may help to put this problem in perspective. Since 1945, there has been a fairly steady increase in undergraduate and graduate enrollments in engineering schools in this country, with some fluctuations, of course. Since the mid-1970s, there has been a quite remarkable increase (about 50%) in the number of engineering bachelor's degrees awarded. But even with this growth, we find that a number of other countries produce far more engineers on a per capita basis than does the United States. This country is now producing about half as many engineers on a per capita basis as Japan, for example, while we are producing seven times as many accountants and twenty times as many lawyers per capita!

While we have seen an increase in engineering enrollments in the recent past, this trend is unlikely to continue far into the future, given the projected decreases in the college-age population in this country. Over the next ten to fifteen years, we can expect a decline of about 25 percent in the number of eighteen-year-olds. In New England, the projected decline is even greater, with a drop of roughly one-third expected over the next two decades. Not only is the number declining, but I would suggest that there is a parallel decline in the quality of high school preparation for eventual careers in science and engineering. This can be seen in such national indicators as the fifteen-year decline in college board scores, in the fact that a significant number of high schools do not offer more than one year of mathematics or more than one year of science, or that, in 1982, only one high school graduate in six had studied more than one year of either mathematics or science. This state of preparation simply removes science, engineering, or any technically based career as a possibility for those young people.

At MIT, these declines in population and preparation are not yet felt, but there is a disturbing trend emerging in the pattern of engineering education and careers beyond the undergraduate years. The number of undergraduate students enrolled in engineering at MIT has doubled over the last decade, while the number of engineering graduate students has increased by a quarter over this same period. At the same time, there has been a modest decline in the percentage of our bachelor's-level engineering students who go to graduate school. But, more importantly, the number who go directly into industry has roughly doubled in the past dacade – a shift that is related largely to the economic facts – salaries.

Nationally, there has been an overall decline in the number of doctoral degrees awarded in engineering, from 3,500 in 1970 to 2,500 in 1980. And while the number of doctorates awarded annually to foreign students has remained roughly the same (about 1,000) over the past decade, the number awarded to U.S. citizens has dropped by half during this same period. This decline in doctoral degrees, particularly those awarded to U.S. citizens, makes it difficult to maintain, let alone increase, the size of engineering faculties. Today, there are about 25,000 faculty positions in engineering around the country, about 10 percent of which are vacant. The percentage of vacancies varies, of course, from field to field, ranging from 4 percent in aeronautics to 16 percent in computer science.

A major factor behind the difficulty in filling faculty positions relates to salaries. In 1978, for example, there was a modest gap between the annual salary for an assistant professor in engineering at MIT and a recent Ph.D. entering industry. By 1981, that gap had grown to 20 percent. We reduced, but did not eliminate, that gap in the past academic year by increasing entry-level salaries for assistant professors in engineering by about 20 percent. Trying to close the gap is a little like chasing a moving target because salaries are rising briskly in industry as well. But I think we must reduce that gap if we are going to compete successfully for the ablest people for our engineering faculties.

Certainly salaries are not the only factors influencing career choices in engineering. Intellectual interests and preferences in professional lifestyles also play important roles in career decisions. Nonetheless, the recent large differences between academic and industrial salaries has made it difficult for engineering schools to attract and keep outstanding faculty members.

Economics make a big difference at every decision point in the system. That economic difference discourages young people at the bachelor's level to continue on for the master's degree, and means, too, that fewer still continue on from the master's degree into the doctoral program. And, ultimately, these pressures mean that fewer people opt for academic careers. At every point along the way, there is an economic factor working against the interests of engineering schools in preserving and enhancing their faculties.

There are also "quality of life" considerations affecting academic careers. Enrollments have risen at a time when faculty sizes have not increased, which means that teaching loads and other demands on engineering faculty are very high. In addition, the environment for doing high-quality research has deteriorated steadily over the past decade, as laboratory instrumentation has aged and federal support for research facilities has declined. The average age of instrumentation in universities is twice as old as that in industry — an enormously important difference during a time when the capacity of instrumentation is being greatly extended by the incorporation of microelectronics, microprocessors, and other modern technologies. Over the past twenty years, federal support for research facilities and instrumentation has declined dramatically (by a factor of five in nominal dollars and a factor of ten or more in constant dollars). Essentially, the federal government has disengaged from providing resources to universities for academic plant in support of the research and development enterprise. And that disengagement has been a major factor in the decline in the quality of life in the university setting.

For many new doctoral recipients, this means that if they want to work at the forefront of research, they will be better supported, in terms of facilities, instrumentation, and finances, in the industrial setting rather than the university setting. This does not bode well for engineering education or for the future of research universities. If we in the universities are to educate students for engineering careers in industry or academia, we must be able to both renew and enlarge our engineering faculties. Improved faculty salaries and improved professional environments will help, but real changes in the patterns of study and employment that have developed over the past decade are necessary, and they will take some time to change. Speeding up the process will be difficult at best, but it seems to me that the nation must make a concerted start at this problem in order to ensure the continued high quality of our engineering enterprise — an endeavor in which universities and industry, indeed the society as a whole, have a common stake.

INDUSTRY AND UNIVERSITIES

The Case for a Joint Research Effort In the Semiconductor Industry

by
Erich Bloch
Vice President, Technical Personnel Development
IBM Corporation

The 1980s will see an increasing penetration of high technology into industry, business, and the day-to-day lives of people around the globe.

Semiconductor and computer technologies are predominant in affecting these developments.

The pervasiveness of these effects can be traced to four highly visible reasons. The first is that semiconductors and computers bring with them immediate cost reductions and quality improvements to manufacturing industries, significant productivity increases for the service sector, and great improvements in the efficiency and effectiveness of the engineer, scientist, and administrator. Second, this new electronics technology and the products around which it is built represent a growing industry sector that contributes directly to the wealth of nations while replacing older declining industries. The third characteristic is analogous with the role of steel and cars earlier in the century. Semiconductors and computers are used as the building blocks for new industries such as telecommunication products, video recorders, home computers, and calculators. Finally, semiconductors and computers make possible the information revolution that permits the economical incorporation of information and decision capabilities into many parts of our daily lives. Using information via computers is similar to packaging and distributing power in the form of engines and motors over the last 100 years.

The two technologies are so highly interdependent that cause and effect can no longer be differentiated. Progress in one depends on progress in the other. Thus, computer architecture and functional performance improvements can only occur with more reliable and speedier semiconductor devices; and the latter in turn can only be made more sophisticated with more highly developed computers in development and manufacturing functions of the semiconductor industry.

Increase in worldwide competition

Worldwide competition in semiconductors and computers is increasing. Both industrial and developing countries recognize the fundamental importance these two industries can play in sustaining healthy domestic economies.

Because of government initiatives, especially in Japan, France, Germany, and England, the technological knowledge base is spreading worldwide. In specific sectors such as memory devices, calculators and displays, the United States has lost its preeminence. For example:

- While U.S. industry supplied 100 percent of its memory chip needs ten years ago and over 90 percent by 1975, this fell to 60 percent by 1980. By 1983 it is estimated that the United States will supply 30 percent of the 64K market while Japan will supply 70 percent.

- The number of papers presented by American-based authors in the prestigious yearly International Solid State Circuit Conference(ISSC) has decreased from 78 percent in 1971 to 55 percent in 1981. Japan's participation increased from 5 percent to 30 percent during the same time interval.
- In U.S.-registered semiconductor patents,the number issued to United States companies has declined in ten years from 570 to 430; the share of Japanese companies increased from 85 to 190.
- The global U.S. market share for semiconductors declined from 70 percent to 60 percent between the early 1970s and 1981, while Japan's share in the same timeframe has risen to 26 percent. The European Economic Community's share has declined slightly to 10.5 percent, while other nations are at a miniscule 3.5 percent.

While America remains strong in computers, the above-mentioned changes and indicators point to the increasing success of countries like Japan in high technology industries such as semiconductors. Such advances demonstrate clearly Japanese desires to participate in and possibly dominate the computer industry. This is also underlined by the initiative of Japan and what has been described in this book as the fifth generation computer project.

Economic environment

Economic and political stresses have ominous implications for the competitive positioning of U.S. companies in the world market.

Since 1978, U.S. inflation (11%) remained higher than Japan's (3.6%); or Germany's (4%). Capital investment in the semiconductor industry, while increasing to almost twenty cents of every sales dollar, is burdened by financing charges twice as high as those in Japan in 1980 and 1981. And because Japanese companies do not rely on equity financing but rather on loans from institutions with an equity position in the enterprise, their after-tax profits can be left at rates as low at 2 percent, provided the market share is increasing. Such circumstances are impossible for American corporations.

Import duties are uneven. While Japan's rate for semiconductors has finally been lowered to equal the U.S. level, Europe is still four times as high. Coupled with buy-local national policies in high technology areas, the marketing difficulties for U.S. companies are great in such fields as telecommunications. Additional regulatory constraints are barriers to entry that contrast strongly with the laissez-faire policy that otherwise prevails in the United States.

The Search for solutions

The United States can only remain competitive by addressing two major problem areas: education and research. On neither count is the nation keeping up with its competitors. U.S. R&D spending has been dropping while it doubled in Japan and Germany during the same fifteen years; if American defense R&D spending is discounted, the differences are magnified. It is incumbent on industry, together with the academic community, to take the necessary countermeasures.

The Semiconductor Research Cooperative (SRC)

The SRC (described in Chapter 6) probably could not have come to life a few years ago. Yet today it is accepted as an idea whose time has come. One reason for this change in view is the realization that the semiconductor industry is under severe and continuous pressure from foreign competition. Having become a basic industry of the country, its viability is essential to the well-being of the national economy. The need for defense support alone would justify this position, but the industrial requirements cannot be minimized. As we look to the future, a lead in semiconductor research today will be the determinant of market performance tomorrow.

Not only has the U.S. research effort been decreasing during the last few years, but costs of doing research have been escalating. Inflation, rising manpower costs, and the need for increasingly sophisticated tools to handle complex technologies are to blame. While some universities and some regional institutes have recognized the importance of semiconductor studies, the early obsolescence of equipment and constant need for new capital infusions are major problems.

These conditions, coupled with a reduction in government spending, make industry support essential. Tools and equipment must be made available. By sharing such resources, the immense costs involved can become more affordable. It is with these conditions in mind that the (SRC) was founded, with a long-term view of the solutions needed.

The main goal of the SRC is to increase the level of focused research by the U.S. semiconductor industry. Its focus is both on the five-to-ten-year efforts that require long gestation, and on the shorter three-to-five-year research projects that will yield tangible products. The direct aim, however, is not at advanced technology or specific product developments but rather on cooperative research that can then be turned into technology and new products by participating companies. Cooperative research does not preclude vigorous competition in the marketplace.

In addition, the SRC will devote substantial new amounts of member company funds to pure research. This should increase understanding of basic concepts that lead to product development and accelerate the applications timetable.

The SRC will assist in adding to the quantity and quality of degreed professionals. This will be achieved by investing the major portion of the research funds in universities and not-for-profit institutes, thereby funding new research positions and attracting new talent to these endeavors.

Finally, the SRC will invest its funds in upgrading and modernizing the tools and equipment essential for semiconductor research. The unacceptable current condition of equipment deficiencies in universities has to be altered to ensure work at the forefront of this industry's technologies.

To achieve these goals the SRC has sought participating companies that are not only semiconductor manufacturers but also major and leading-edge users: computer firms, instrument and equipment companies, defense-oriented as well as consumer product-minded corporations. What these companies have in common is a strong dependence on semiconductors. For them, continued competitive advantage in a global market means reliance on a dependable and vigorous industry.

A vital question

Why should individual companies not merely increase their own internal research efforts or interact one-to-one with selected universities? While such an approach is feasible, a joint effort can bring to bear the necessary critical mass to attack a crucial research area. A joint effort can avoid an overlap of endeavors that is all too often the case when individual companies pursue their own insular research agenda.

Cooperative research has proven beneficial for other U.S. industries such as textiles, electrical and gas utilities, and some sectors of the chemical industry – with highly applauded results. And of course one notes similar and very successful undertakings in Japan. In the latter case, research is carried out under the auspices, monetary support, and control of the government.

Getting started

While at the time of this writing (April 1982) plans have not be fully firmed up, the SRC is contemplated as a nonprofit organization operating as a subsidiary of the Semiconductor Industry Association (SIA). It will be governed by a board of directors, a third of whom will be elected by, and will be menbers of, the SIA board; the other members will come from participating industry and academia. The SRC will operate without government support or participation.

An executive director will report to the board with the responsibility of formulating research programs, entering into activities with universities and other institutions, and monitoring and disseminating information. The executive director will be assisted by a small group of technical people from member companies.

Member fees will be based on sales volume. Preliminary goals (while not confirmed) are for a starting fund of $5 to $7 million in 1982, and $10 to $15 million in 1983, hopefully increasing thereafter with the number of participants and the growth of the industry. The 1983 fund alone will be 25 percent larger than the 1980 industry research budgets – a not insignificant increase. Early estimates forecast a doubling of the yearly support by the semiconductor and computer industry to university research.

Membership will be open to all companies that manufacture semiconductors in the United States. The SRC and participating companies will own patent rights and other intellectual property, which can be licensed to nonmembers for an appropriate fee.

Closing thoughts

The U.S. semiconductor industry has reached a critical crossroad. At no time in its thirty-year-history has it been assailed by as much competition and as many economic problems. A well-directed research undertaking could significantly alter this industry's self-confidence and its reputation both at home and abroad. The SRC will demonstrate in a most visible way that the semiconductor industry and its major leading-edge users are willing to work together to preserve and enhance its independence and long-term vitality.

THE SIGNIFICANCE OF NEW UNIVERSITY/ INDUSTRY RELATIONS

by

Robert M. Hexter
University of Minnesota

We are embarked upon a major experiment. We have undertaken to determine if American universities and industry can significantly pool their talents and resources to work together toward common goals: to create new knowledge and to train more people better.

It is not going to be easy. For too many years each of the participants has viewed the other's motives, values, and goals with suspicion, if not disdain. To businessman Andrew Carnegie, a classical education better fitted one "for life upon another planet." To intellectual Thorstein Veblen, eagerness to please a financial benefactor was a "fatal step." Extreme views such as these are still commonplace.

Universities and industry need each other. The conditions of mankind on this planet do not permit a return to the relatively quiet life of the early part of this century. Like it or not, the world has grown into a densely populated, resource-hungry community struggling to govern and control itself. We can use all the help we can get by increasing the levels of both technology and education in our lives. Indeed, many of our present problems derive from the insufficient development of new technologies reflected in sliding R&D expenditures during the past fifteen years.[1]

Coupling these increasing needs with current declines in federal and state support for universities, there is every reason for universities and industry to look to each other for help.

Can it work? I believe not only that it can but that it must, because a new synergism is becoming critical to the vitality of both university and industry research.

Creativity in basic research, always a will-o'-the-wisp to American industry, is under further pressure in many companies. Notable counter-examples of excellence – Bell Telephone Laboratories, IBM Watson Laboratories, or the once-great Shell Development Laboratories[2] – cannot obscure the "Annual Report" problem: the misperceived need to demonstrate "profit" in each year's basic research. The problem is not a lack of good research personnel; rather, it derives from management structures unable to transform a research idea into its realization. Creativity thrives where individuals can make their own decisions about the direction of their work and lives. People want to "sign up," not to be assigned. The annual report attitude inherent in much of today's industrial environment thwarts such creativity.

At the same time, the cracks and flaws in the university research edifice grow more apparent. A system that has produced so much new knowledge in this century alone – from quantum mechanics to gene splicing – has now become costly and overloaded, with some individuals "tenuring-in" and coasting on their all-too-well-constructed research flywheels. At the same time, we are faced with the potentially catastrophic situation of having fewer and fewer new faculty to teach a larger and larger undergraduate student body, as industry is offering such extraordinary salaries to baccalaureates that the graduate student population is becoming dangerously small. With a tradition for slow and carefully considered change, universities have failed to keep pace with a world that is changing at a pace faster than traditional academic structures can absorb.

The obvious synergism needed in the face of these dilemmas is a mechanism for closer collaboration and exchange. In the past, company-supported university research has been done at more than arm's length. The university attitude has been: "Give us a grant and once a year we will give you a report." We can no longer work this way. There is a need for true collaboration. Company research personnel, for example, need to be on the campuses – perhaps on a sabbatical leave basis. They should not only work in collaboration with professors but also serve as visiting faculty members. Universities should develop the practice of involving industrial research personnel as co-mentors of graduate students, to share the responsibilities and rewards of guiding students toward advanced degrees.

In a truly collaborative process, technology transfer will really take place, and more will be achieved in both directions than by any other mechanism proposed so far. Furthermore, in providing this closer link, industry will gain more direct access to and evaluation of a new supply of students. Of greatest importance, faculty will be exposed to the long range needs of industry for new technologies and breakthroughs, and to the extent these are interesting and challenging, university researchers will turn their attention to them. While exchanges of personnel between universities and industry have been tried before, they have not been attempted as part of a sustained collaborative effort.

One way to implement this kind of interaction would be to couple company grants for basic university research to a provision for the release of, say, 25 percent of the time of a corporate researcher to work collaboratively with faculty members. The dividends of such an investment – in the form of technology transfer, in providing for a greater access to a supply of graduates, and most of all in maximizing the creativity of all concerned – will be many times that afforded by our present, separate ways.

This kind of working relationship is being fostered by the Center for Microelectronics and Information Sciences (MEIS) at the University of Minnesota. MEIS is presently sponsored by four major corporations – Control Data, Honeywell, Sperry Univac, and 3M. Consider just one example of collaboration, designed to support experimentation with the very ways that the university carries out instruction. In place of the large lecture format, students will increasingly be able to benefit from an ideal one-on-one teacher-student tutorial mode by the use of microcomputers networked to large mainframes. The development of such a system in a variety of disciplines is probably beyond the capabilities of even a large university, particularly at this time. But consider also the intellectual power of such a consortium. The products of this collaboration have immense potential to teach more students better, as well as to address the continuing challenge of retraining engineers to continue the advances of technology.

This is the major experiment we are embarked upon. The purpose is not simply to facilitate the university-industry interface; it is to reestablish the storehouse of basic research needed for the technology of the 1990s and beyond. While no single way is best in initiating this collaboration, MEIS represents a significant point on a national learning curve. A variety of experimental forms are called for. The question is not, "Can we do it?," but, "How soon do we begin?"

INDUSTRY–UNIVERSITY PARTNERSHIP IN THE HIGH TECHNOLOGY AGE

by

Kenneth G. Ryder
President, Northeastern University

The Problem

The progress of civilization is metaphorically and perceptively described by Alvin Toffler as a progression of waves. The first wave, representing agricultural civilization and lasting thousands of years, preceded a second wave of industrial civilization. Now gaining hold is a third wave that includes the newer era of electronics, computers, and information processing – more concisely labeled by the term High Technology.

Many of the problems we now experience are portrayed as the turbulence and conflicts resulting from the collision of the waning industrial second wave and the rising third wave. The wave analogy is revealing. Indeed, current imbalances in the use and distribution of human resources in industry and academia can be ascribed to a failure to fully appreciate the changing tide from machine-intensive industry to an information-intensive economy.

My view of high technology is from the perspective of a university with a predominantly professional orientation. While other institutions share similar concerns that all segments of society have open access to higher education, Northeastern University is unique in its extensive commitment to cooperative education and to comprehensive adult and continuing education activities to achieve this goal.

Cooperative Education

Classical models of education, traceable to European medieval centers of learning for clerics and members of an aristocratic elite and surviving through to the present, are no longer appropriate. The complex needs of contemporary society, in which a large proportion of the population requires extensive training, call for a new definition of responsibility to educate. It is in this light that the cooperative education philosophy — integrating academic study with career-related work experience — is seen by us as part of the wave of the future in higher education. It is clearly in answer to a technology-driven society such as ours that cooperative education provides a responsive link between the university and the rapidly evolving world of industry and commerce.

Cooperative education bridges the gap between formal study and on-the-job experience. It is called "cooperative" because both industry and academia must combine interests to create a cohesive educational program. Students alternate study periods with career-related work which is compensated and supervised by employers. This interrelationship is carefully planned to achieve optimum educational results.

Northeastern University's experience with "coop" programs started in 1909 with engineering. Since then the concept has expanded to include a variety of other professional fields. The benefits have shown over time to be of equal value to undergraduate and graduate students. Every year, the University currently places 11,000 students with about 2,500 employers throughout the United States and several foreign nations. Collective earnings from this on-the-job learning total about $60 million a year and are used to assist in meeting their educational expenses. More importantly, it is the invaluable exposure that comes from working alongside practicing professionals that is brought back to the campus by student participants. This said, it is easy to understand the enthusiastic commitment of Northeastern University in encouraging a close partnership between industry and all institutions of higher learning. This seems especially pertinent as we seek to address the serious shortage of trained manpower looming over the coming decade.

Critical shortages of manpower in high technology fields are well documented. Where differences of opinion do arise on this question, they pertain more to the magnitude of the shortage and its durability than to its acknowledged existence. High technology industries are clearly dependent on higher education for the development of the professional manpower they need. Many barriers to the fulfilment of those needs can be identified: shortages of faculty, obsolete facilities, declining federal support to education, and demographic changes in the distribution of the population. These factors bear directly on the quality of instruction and the quality of graduates who ultimately are employed by industry.

Continuing Adult Education

Discussions focused on meeting the forecasted needs of industry are often limited to ways and means of expanding the undergraduate pipeline of candidates. Another body of opinion argues that significant manpower increases can be achieved by reversing the rapid obsolescence of practicing professionals. A coherent and realistic response would seek to encompass both of these approaches – and would differentiate between short-range and long-range solutions.

The current condition of American higher education is unprecedented. While an urgent need for people with specific skills is not new, what is new is the magnitude of the need, the power of domestic and international market forces driving it, and the abdication of federal support in arriving at a solution. These circumstances are unparalleled. Thus, it becomes clear that conventional academic actions will not yield the desired outcomes. Only carefully conceived, bold moves predicated on clearly defined objectives are likely to provide a reliable manpower base for America's high technology industries of the third wave.

As we see it, these objectives should serve: (1) to establish enduring mechanisms for multifaceted links with industry; (2) to create a flow of new high-quality technical manpower educated with sufficient breadth to adapt to changing needs; and (3) to institute flexible means for updating practicing engineers throughout their professional careers. The strategies we advocate to meet these objectives are based on our judgment that:

- all available evidence supports the contention that high technology developments will further dominate industrial growth;
- close links between industry and academia are essential to meet the needs of high technology industry;
- cooperation among compatibly focused and situated universities will enhance the effectiveness of academic responses;
- engineering would benefit from a structure analogous to that in other professions (e.g. medicine and law) where successful practitioners provide instruction and are paid at market rates in their field; and
- a significant portion of engineering education should be conducted in a practice-oriented setting.

Reconciliation of different time horizons is a key to a successful link between industry and academia. The former, pressed by intense competition, seeks rapid conversion of scientific and technological developments into marketable products or services. The latter is restrained by an apparent inertia that accompanies a desire to achieve slow and steady educational proficiency in new fields. In addition, the cyclical effects of demand and supply for technical manpower run counter to quick responses by academia in meeting needs of the market. There is little reason to believe that the pattern of peaks and valleys on the demand side will change in the future, so that the real challenge is to achieve a better match in the temporal relationship between supply and demand.

Today's shortages require that existing pools of trained manpower be tapped to yield new proficiencies after relatively short training periods. A clear identification of needs will allow the appropriate training programs to be designed. These should be designed not only to fit specific occupations but to account for prior training of students. These programs should be based on flexible schedules. They should be intensive. They should lay the groundwork for future updating or adaptation to related activities. Special counseling should be added to account for various degrees of aptitude students might have for technical work.

For example, Northeastern University instituted several programs during the past two years. A program in technical writing was created to train people for entry-level positions. A master's degree program in English with an option in technical writing was also initiated. In addition, we offer a one-year course of study for college-educated individuals combining intensive training in technical writing and computer science with a six-month-long paid internship period in industry. A Computer Systems Specialist program for students with varied backgrounds provides for entry-level programming employment. It includes a cooperative work assignment with an optional final phase of full-time employment, part-time study.

Drawing on an underutilized segment of the population in technical fields are graduate programs for Women in Engineering and Women in Science. The curriculum in both cases is designed to update previously acquired knowledge and to refocus energies toward new career opportunities. These programs – as well as many efforts under way elsewhere – exemplify a variety of retraining models allowing participants to redirect careers to fields in demand by industry.

Adult and continuing education approaches hold great promise for providing the technical training sought by high technology industries. Northeastern University is known for its tradition of leadership in these fields. Over 30,000 students enrolled in its programs in 1981-1982.

The long-range goal of ensuring a continuing, high quality flow of technical manpower suggests several other directions. With engineering schools already operating at full or near full capacity, particular attention must be focused on those factors that either inhibit expansion or threaten a decline in quantity and quality of graduating engineers. Faculty shortages and equipment obsolescence are the most visible problem areas for engineering departments.

Industry and academia both tap the same finite pool of engineering talent. It is now abundantly clear that the competitive advantage in attracting such people is with industry, where high salaries and other benefits cannot be matched by academia. A solution to this imbalance might be found by the partial blurring of the distinction between the industrial and the academic engineer.

We propose that a widespread utilization of part-time, adjunct faculty from the industry side be matched by an expansion of consulting and cooperative research arrangements for faculty members on the academic side. At the same time, imaginative applications of new teaching technologies should be aggressively pursued to achieve a leverage of instructional efforts.

This approach will forge stronger industry/academic links, will provide instruction by practicing professionals, and will help to update and maintain the quality of academic faculty members. Of special importance, again, is the underlying premise that the intrinsic commonality of the pool of people available for teaching must be recognized and reflected in comparable compensation. The professional overlap and the diminution in income differences, if achieved, can restore the attractiveness of graduate study and academic careers.

Major developments in high technology are coming at an ever-increasing pace. It is no longer surprising to find that the time interval between new generations of technology is of the same magnitude as the time required to educate an engineer. As a result, the effectiveness of the practicing engineer can rapidly diminish if steps are not taken to prevent it. Special efforts to offset professional obsolescence will help to dampen the demand for new talent and thereby alleviate the shortage of specialized manpower. We advocate the institution of carefully planned, individualized programs of continuing education for these professionals.

A systematic program linking work to study – in effect an advanced level of cooperative education – is of critical importance. Flexibility is the key to its success. An innovative example is the Early Bird graduate engineering program at Santa Clara University in California's Silicon Valley, which holds special classes between seven and nine o'clock in the morning. Into such a program one can incorporate the adjunct teaching activities discussed earlier with a learning experience for the industrial instructor.

Industry-University Cooperation

Access to up-to-date equipment is essential for students in rapidly evolving high technology fields. Yet several years of reduced federal funding for equipment in universities is creating serious problem areas. While industry has traditionally been generous in helping, new support will be needed. Tax incentives must be extended to provide for equipment donations for educational purposes beyond those directly associated with research.

Because the needs are so broadly based on the part of universities, the incentive for collaboration between institutions in dealing with industry is great. Yet in the past a strong tradition of institutional autonomy has tended to inhibit fruitful cooperation and encourage rivalry. Such traditions are rapidly changing in this age of high technology.

In Massachusetts, for example, an informal consortium of eight technical universities has assembled at the presidential level at my initiative. The purpose of this grouping is to explore ways of avoiding wasteful duplication, to exchange information, and to identify areas in which direct collaboration would be desirable and feasible. Additionally, the group will deal with such concerns as the formulating of educational programs appropriate for technological developments over the next decade, and the search for strategies to update professional skills and to gain state government support for joint ventures in high techology fields. The consortium is currently working with government and industry leaders to plan a microelectronics laboratory that could be shared by all universities.

The highly effective association of companies known as the Massachusetts High Technology Council and the consortium of universities might well serve as models for productive industrial and academic interaction in other regions of the country.

My view is that universities must become more adept at responding to the needs of a rapidly evolving high technology-based society. Universities must break away from traditional, rigidly compartmentalized programs, and become attuned to the pace of technological change. Universities must learn to work together, with industry, and with government to accomplish commonly defined goals.

Northeastern University is striving to adhere to these principles. Through our planning efforts, three areas have been identified as areas of university strength: technology, management, and urban affairs. These are areas in which we have the potential to contribute to progress in the decades ahead. An emphasis on cooperative education and adult education will play a central role in our plans. Our experience, gathered over a seventy-year period, will be willingly shared with others.

Cooperative education has proven particularly effective as a bridge between industry and academia. I believe that Northeastern University stands as a model non-traditional center of higher learning for others to emulate here and abroad.

ELECTRONICS, EMPLOYMENT, AND KNOWLEDGE SOCIETY

by

Jean Saint-Geours
Special Advisor to the Prime Minister of France

The French government is proceeding with a clear sense of mission concerning "informatique" and France's role in the global economy. At present in France, there is consensus between the political and economic leadership to pursue with determination a strategy of developing the nation's potential in electronics, computers, and telecommunications. Few people are opposed to intensive applications of informatics technology – robotics, office automation, and "telematique" – to industrial production processes, to the service industries, and more generally, to all that is affected by information and communication.

It has been firmly established that French participation in a global economy based on free exchange is essential. Any slowing down of its technological progress is out of question for fear of losing competitiveness and ultimately regressing economically. The socialist government is already taking direct action and reallocating budgets for more support – whether it be to the computer and peripherals industry, to pure research and related development, to the expansion of software applications, or to the creation of experimental systems technologies.

To a higher degree than in the United States, the private sector in France has not had sufficient resources to take on the financial risks inherent in these activities. This is a justification for the socialist government to nationalize the largest enterprises responsible for technological and industrial advances. This means that the government is committing itself to accelerate advances deemed essential to the national interest. The present government has not cancelled prior French policies in informatics, but rather intensified them.

The socialist government aims to modernize industries such as machine tools, textiles and garments, and forest products to become more competitive in international markets. The automation of production is one of the most important means for modernization. As a result, we will see a diminution, now and even more so in the future, in the amount of labor needed for a unit of production. This process is indispensable in limiting foreign competition, particularly from the Third World.

Jobs and Informatics

These ambitions are nevertheless confronted by a powerful social and psychological barrier: the labor force is underemployed, and unemployment is growing daily (more than 10 million in the European Economic Community, more than 2 million in France at the end of 1981). Informatics tends to create less new employment in its direct production or in new information services than it eliminates through its application in other sectors of the economy. This is hardly unique to France.

Pressures to automate are strong. Robotics is actively brought onto the assembly line not only because of the vital productivity gains but also because of other benefits, such as improvements in the work environment and the substitution for hard-to-find specialists. The same is true for office automation, with the added benefit of upgrading the analysis, circulation, and quality of information. Database automation and the possibilities of real-time control of physical or financial operations associated with systems of distribution, credit checking, or other services will lead to a new level of competence and know-how in whole sectors of the economy: for example, in libraries, publishing, and banking operations. More generally, telematics offers an opportunity to eliminate certain activities associated with the physical movement of people and goods – for example, transportation and hotels.

Surely one should not overdramatize these difficulties. Robotics, telematics, and office automation are indeed generators of new jobs. In the largest sense, one can expect many new developments created by the interaction between the electronic technologies and new needs or desires associated with them. We know well the "complicity" that occurs between the imagination of the supplier and the curiosity of the client, which contributes to the increasing proliferation of diversified services.

It is also important to note the important role of electronics on the further development of other new sectors – nuclear power plants, aeronautics and space research (launchers and satellites), telecommunications for business, science, and entertainment, and national defense. For decades, these have been promoted jointly by the public and private sectors.

The hour is at hand to adopt strategies that will accelerate and exploit these new technologies to their fullest. The bottom line to all these actions is that there is a risk of unemployment. The numbers of unemployed may tend to grow dramatically in France. However, to repeat, this should not imply putting the brakes on the progress of informatics. The solutions to unemployment that is caused by technology cannot be found by stopping progress but must be found elsewhere.

The New Society

The outlines of a "new society" are being drawn on the horizon of the future, a society in which the "four seasons of life" may become differently distributed. Learning, work, creation, and leisure may become sequenced in unconventional ways. A socialist France will strive to spread the employment opportunities to create better conditions for the constructive use of free time. An information or knowledge society also has other implications for changes in education and work.

Time devoted to learning, education, and apprenticeship will have to be expanded. This simply reflects higher levels of knowledge required to invent, organize, and control the production of goods and services. It also reflects the increasing specialization of professions. At several points during their lifetimes, chief executive officers, managers, workers, and employees will have to bring their competences up to date or even change fields completely. This fact creates new problems in defining "general education" or, more specifically, its ability to adapt to constant changes in the economic environment. As a result, the content and methods of education will have to be restructured to remove the barriers between formal education and the needs of the individual and society.

In contrast to the increased time devoted to education, time devoted to work will be reduced through automation. This may provide an opportunity to further redistribute worktime to absorb the unemployed. The current objective is to reach an average work week of thirty-five hours by 1985. Ultimately, the work methods themselves will change. It will no longer be necessary to have all workers gathered at one place at one time. Their tasks can be decentralized and sequenced – a freer setting but possibly less convivial.

The introduction of computerization runs the risk of severely impoverishing our culture. One must anticipate this impoverishment, but it cannot be legislated away. To introduce new cultural tools is a difficult task that will necessitate new methods of teaching. Here, the new media can play a very significant role, if intelligently used. What magic hand could bring these new technologies to life instead of turning them into mechanical professors?

The ground rules of a new civilization are emerging. Provided with so many resources, will people be able to use them wisely? A fundamental aspect of this problem is already surfacing in France: finding an equilibrium between old traditions that need to be preserved, and those that must be broken and overthrown. Throughout French history, there has been a dialectic of conservatism and revolution. But the advent of informatics undoubtedly constitutes an unprecedented break with the past. Our political, administrative, economic, and social institutions need to make a response, but they are rigid as well as powerful. It is important that they understand the evolution under way, both adapting to it and helping in the adaptation. The chances seem optimistic that it will indeed happen this way. Otherwise, like that inscrutable dictator of a Gabriel Garcia Marquez novel, we will wake up one day to discover that our institutions have long been dead without the populace ever realizing it.

NOTES

INTRODUCTION

1. Thomas Kuhn, *The Structure of Scientific Revolutions*, Chicago: The University of Chicago Press, 1962, 1970.
2. Jay Forrester, "Innovation and the Economic Long Wave," *Management Review*, June, 1979
3. Robert Hayes, "Managing Our Way to Decline," *Harvard Business Review*, July/August, 1980.

CHAPTER 1

1. The full comparison between American and Japanese cars is as follows:

DETROIT VERSUS JAPAN: Questions of Quality

1980 American Models	Mechanical and Body Ratings[a]	
Buick	9.4	9.3
Chevrolet	8.9	8.4
Dodge	10.0	10.0
Ford	9.2	7.2
Lincoln	8.4	8.1
Oldsmobile	9.3	8.4
1980 Japanese Models	**Mechanical and Body Ratings**	
Datsun	10.8	15.3
Honda	11.1	16.0
Mazda	12.7	17.5
Toyota	12.4	16.9

Source: "Industrial Competition," *Harvard Business Review*, September/October 1981.

[a]Scale = 1 to 20

2. "Kawasaki to Sell M.T.A. Subway Cars," *New York Times*, March 18, 1982.
3. Report to the Ministry of Industry (France), Interex Associates, Lincoln, Massachusetts, 1980.
4. Part of the reason is that many French companies do not have in-house data processing staffs; they contract out many of the services needed for day-to-day operations. This tends to inflate the total sales picture in comparison with the United States.
5. *Le Monde Informatique*, October 5, 1981.
6. *Le Monde Informatique*, February 22, 1982.
7. Peter Drucker, *Managing in Turbulent Times*, New York: Harper & Row, 1980, p. 169.

8. Lester Thurow, "The Productivity Problem", *Technology Review*, November/December 1980, p. 40.

9. Lester Thurow, "Other Countries are as Smart as We Are", *New York Times*, April 5, 1981.

CHAPTER 2

1. According to the Harvard Program on Information Resources Policy, the total information industry in 1975 had gross revenues of nearly $700 billion. Of these, $100 billion (15%) were in telecommunications, electronic components, radio and TV, and computer systems, software, and services. See Note 3 for further details.

2. Fritz Machlup, *Knowledge: Its Creation, Distribution, and Economic Significance*, vol. 1: *Knowledge and Knowledge Production*, Princeton, N.J.: Princeton University Press, 1980.

3. Marc Uri Porat, *The Information Economy*, 9 vols., Washington, D.C.: Government Printing Office.

On the question as to what is counted as an "information industry," the Program on Information Resources Policy at Harvard University lists the following:

WHICH ARE THE INFORMATION INDUSTRIES?

Industry	**Gross Revenue ($ billions 1975)[a]**
Telephone, telegraph, satellite, and mobile radio	$ 35.1
Postal services (public & private)	11.6
Paper, pulp & board; photo equipment & supplies	23.1
Radio & TV (equipment and broadcasting)	23.3
Computer systems, software, services	22.6
Electronics components	20.3
Motion pictures, sports, theaters	9.1
Newspapers, periodicals, books, and publishing	19.7
Schools, libraries, and research	148.1
Advertising	10.0
Business consulting	1.8
Brokerage services, banking, insurance & other financial	350.6
Government: national intelligence, social security administration, county agents	12.7
Legal services	14.8

[a] Where 1975 data is not available, the next earliest year is used.

Source: adapted from data reprinted in R. Hamrin, *Managing Growth in the 1980s*, New York: Praeger, 1980.

4. Arnold Toynbee, *Mankind and Mother Earth, A Narrative History of the World*, London: Oxford University Press, 1976.

5. Alvin Toffler, *The Third Wave*, New York: William W. Morrow, 1980.

6. Porat, *The Information Economy*, vol. 1, p. 4. Porat defines two types of activities: "one in the primary information sector where information is exchanged as a commodity, and one in the secondary information sector where information is embedded in some other good or service and not explicitly exchanged." On the question as to whether a tractor is part of the agricultural or industrial economy, Porat says a tractor is "obviously part of the food activity."

7. Further classifications would include: (1) electronics (but not "electrical" – for example, spark plugs); (2) information processing: computers (both hardware and software), office equipment (including word processors), peripherals and services; (3) robotics, numerical control machines, and CAD/CAM devices; (4) scientific instruments, including measuring devices and controls; (5) semiconductors; and (6) communications and telecommunications.

8. "Capital Invested Per Employee, 1975," New York: The Conference Board, New York, January 1977. Dollar figures are in constant 1958 dollars.

9. Michael Boretsky, "Trends in U.S. Technology: A Political Economist's View," *American Scientist* 63, January/February, 1975.

10. Ira C. Magaziner & Robert B. Reich, *Minding America's Business*, New York: Harcourt Brace Jovanovich, Publishers, 1982.

11. Ibid.

12. *The Nation*, September 12, 1981

13. Figures are from Business Week Team, *The Reindustrialization of America*, New York: McGraw Hill, 1982, p. 18. They are expressed as a percent of GNP. In absolute dollar terms, only IBM and ATT are high tech companies large enough to be among the top ten:

Top Five Companies in R&D Spending in Millions of Dollars:

1. General Motors	2,224
2. Ford Motor	1,675
3. IBM	1,520
4. ATT	1,338
5. Boeing	767

14. *THE TOP TEN U.S. R&D FIRMS:*
(All Are in the Information Sector)

Company	Sector	R&D as percent of sales
1. Amdahl	computers	15.8%
2. Cray Research	computers	14.5
3. Cordis	electronics	12.3
4. Auto-Trol Technology	computer-aided design	12.0
5. Applied Materials	semiconductor mfg equip	11.7
6. Kulicke & Soffa	semiconductor mfg equip	11.4
7. Intel	semiconductors	11.3
8. Wavetek	instruments	10.9
9. Floating Point Systems	computers	10.8
10. Siliconix	electronics	10.7

Source: *Business Week*, July 6, 1981.

15. John Naisbitt, *The Restructuring of America – High Tech, High Touch*, Lowell, Mass.: Wang Laboratories, 1981.

16. Many other countries have important roles to play. Not only countries like Canada, Australia, West Germany, England, Ireland, Scandinavia, Italy, Austria, Switzerland, Holland, and Spain, but also newly industrializing countries like Brazil, South Korea, Taiwan, Hong Kong, Malaysia, Singapore, Mexico, and Venezuela. Also, the centrally planned economies – especially the U.S.S.R., Czechoslovakia, Hungary, and the G.D.R. – have significant computer industries. A review of all of these is outside the scope of this book. We will focus instead on the challenges from Japan and on the advanced French thinking about international strategy.

17. Simon Nora, *L'Informatisation de la Société*, [The Computerization of Society], Cambridge, Mass.: MIT Press, 1980.

18. *Electronics*, January 13, 1982

19. Dr. Robert Noyce, vice chairman and founder of Intel, notes that the technology is moving faster than expected, and that already bubble memory storage devices are available that hold 10^8 or 10^9 bits of information or more in the requisite size. "You need to raise these numbers by at least one if not two orders of magnitude." (Interview spring 1982).

20. The details of the comparison between the ENIAC and Fairchild F8 are given by Juan Rada in an unpublished paper:

TECHNOLOGY PROFILE: 30 Years of Change

Comparison of ENIAC (1946)
with the Fairchild F8 Microprocessor (1976)

Size	3000 cu.ft.	.011 cu.ft.	300,000 times smaller
Power Consumption	140 KW	2.5 Watts	56,000 times less
Capacitors	10,000	2	5000 times fewer
Mean time to failure	hours	years	10,000 times more reliable
Weight	30 tons	1 lb.	60,000 times lighter

Source: Juan Rada, *The Semiconductor Industry*, unpublished paper.

21. TMA Management Consultants, *The High Technology Enterprise in Massachusetts, report prepared for the Commonwealth of Massachusetts*, October 1979, p. 27.

22. Guenter Friedrichs and Adam Schaff, ed., *Microelectronics and Society: For Better or for Worse*, Report to The Club of Rome, Oxford: Pergamon Press, 1982.

23. Ibid.

24. See, for example, "The Plan for an Information Society: A National Goal Toward the Year 2000," Final Report of the Computerization Committee, Japan Computer Usage Development Institute, May 1972.

25. Boretsky, op. cit.

26. Data General, Internal Corporate Report, November 1981.

27. "Is There Really a Shortage of Engineers?", *New York Times*, March 28, 1982, special insert on high technology.

CHAPTER 3

1. Norihiko Maeda, director of the electronics policy division of the Ministry of International Trade and Industry (MITI) in "La Politique Informatique au Japon," *Informatique Coopération Internationale et Indépendance*, Actes du Colloque International Informatique et Société, vol. 4, Paris: La Documentation Française, 1980.

2. "France Entering the Computer Battle," *New York Times*, April 14, 1967.

3. *Vision of MITI Policies in the 1980s*, Tokyo: MITI.

4. "A Japanese Explains Japan's Business Style," *Across the Board*, February 1981.

5. "The Japanese Invasion: chips now, computers next?," *Electronic Business*, July 1981, pp. 84-89.

6. *Business Week*, December 14, 1981, p. 120.

7. National Science Foundation, *Science & Education for the 1980's & Beyond*, Washington, D.C.: NSF, 1980, p. 59.

8. Ibid.

9. "MITI's Industrial Policy in Japan Survey," *The Economist*, February 23, 1980, p. 27.

10. "March of the Robots, Japan's Machines Race Ahead of America's," *The Wall Street Journal*, November 24, 1981, p. 1.

11. IBM Press Release #022582 and personal correspondence with Erich Bloch.

12. The figures for Japanese government support to computer related research are as follows:

CONCENTRATION IN JAPANESE RESEARCH

Major Government-Supported Research Programs
(in millions of U.S. dollars)

Hardware	**Year**	**Amount**
Voice recognition systems and pattern information processing	1971-80	$104
3.75 series computer and peripheral development	1972-76	290
Integrated circuit (IC) development	1973-74	17
VLSI development (government only)	1976-79	139
VLSI (government and industry jointly)	1976-79	350
Base technology for new era computer	1979-83	112
Scientific processor (super computer)	1981-88	100-148
Fifth generation computer (in planning)	1981-90	400-500
Opto-electrical measurement and control	1979-86	86
Opto-IC development	1981-90	NA
Software		
Software module development	1973-75	14
Software product technology	1976-81	32
IPA agency (program development)	1971-80	52
Leading computer manufacturers	1981-84	150
Independent software houses	1981-84	30
Other: Loans to Japan Electronic Computer Corp. (JECC) & software company	1971-80	2,387

Source: Computer Note, Japan Electronic Computer Corp., and "Japan's Strategy for the 80s," *Business Week*, December 14, 1981.

13. Colloque International Informatique et Société, Paris, September 24-28, 1979.

14. *Paris Match*, January 28, 1967.

15. The French figures for government support of computer related research are as follows:

EXPENDITURES UNDER FRANCE'S PLAN-CALCUL:

	1967-1970	1970-1975
	(In millions of French Francs)	
Subsidies to the informatics industry	480	750
Aid to peripheral and components manufacturers	120	160
Training	—	420
New applications	40	120
Telecommunications	—	256
Research	85	200
TOTAL	725	1,906

Source: J.J. Fabre and T. Moulonguet, *L'Industrie Informatique*, reprinted in J. Rada, *The Semiconductor Industry*, unpublished manuscript, page 157.

16. "Vive La Technologie!," *Economist*, January 30, 1965 p. 456.

17. *Chronicle of Higher Education*, November 4, 1981 page 22.

18. "France Plans World Computer Center," *Washington Post*, January 22, 1982.

19. Jean-Claude Simon, *L'Education et l'Informatisation de la Société*, Report to the President of the Republic, Paris: Fayard Press, 1981.

20. Ambassador Michael B. Smith, "Trade Issues in Advanced Technology," in a speech before the International Information Industry Conference, Quebec, Canada, June 1, 1982.

CHAPTER 4

1. William G. Bowen, President, Princeton University, Annual Report, April 1981; also *New York Times Magazine*, October 18, 1981, p. 98.

2. American Electronics Association (AEA), "A Call for Action to Reduce the Engineering Shortage", Palo Alto, California: AEA, August 1981, p. 2.

3. The figures used here are drawn from the Bureau of Labor Statistics, the American Electronics Association, and from Betty Vetter, "Supply and Demand for Scientists and Engineers," Scientific Manpower Commission, January, 1982. The figures on the size of the electrical engineering workforce are from the National Science Foundation, *U.S. Scientists and Engineers, 1980* (forthcoming). It should be noted that there is considerable uncertainty about many of the statistics.

4. American Electronics Association (AEA), op. cit., p.2.

The AEA demand figures were made by taking as a baseline projections made by the 671 AEA respondents, which represents $77.7 billion annual sales and 500,000 employees. As this is about one-third of the U.S. total, the figures were then multiplied by a factor of three, and compounded annually by 4.98 percent to account for losses due to promotion into management. The demand reflects new growth only and does not include replacement due to retirement, turnover, or death.

The supply figures were made by taking as a baseline numbers of B.S. degrees awarded in 1980. The BS/EE projections were made using a 2.6 percent annual compounded growth rate (ACGR) projected by the U.S. Bureau of Labor. The BS/CS degree figures were reached using a 12.4 percent ACGR, duplicating a pattern of 1977-to-1980 degree increase. All degree projections were reduced by 20 percent to account for the graduates who do not end up taking jobs in engineering.

The widely distributed AEA analysis anticipates electronics industry far outrunning the current capacity of colleges and universities to meet its needs. All else remaining the same, by 1985 the AEA, based on a survey of its members, estimates a supply of new engineers providing only 70,000 new bachelor's graduates in electrical engineering and computer sciences to meet a cumulative need of 199,000. This leaves a gap of 129,000. While there is some contention about the statistical methods used to arrive at these numbers, the dimension of the problem forewarns of significant imbalances.

5. AEA, op. cit., p.3.

6. The numbers of engineering graduates in the United States for the decade 1970 to 1980 are as follows:

NUMBER OF ENGINEERING GRADUATES IN THE UNITED STATES

		Number of Degrees Electrical Engineers		
Year	**B.S.**	**M.S.**	**Ph.D.**	**Total**
1970	11,921	4,150	873	16,944
1971	12,145	4,359	899	17,403
1972	12,430	4,352	850	17,632
1973	11,844	4,151	820	16,815
1974	11,347	3,702	700	15,749
1975	10,277	3,587	673	14,537
1976	9,954	3,782	644	14,380
1977	9,837	3,674	574	14,085
1978	10,702	3,475	524	14,701
1979	12,213	3,335	545	16,093
1980	13,594	3,660	523	17,777
		Computer Scientists		
1970	139	185	34	358
1971	174	250	44	468
1972	359	627	83	1,069
1973	568	589	96	1,253
1974	727	723	83	1,533
1975	599	678	107	1,384
1976	796	718	90	1,604
1977	1,280	802	136	2,218
1978	1,546	986	123	2,655
1979	1,510	1,074	190	2,774
1980	1,816	1,262	159	3,237

		All Engineers		
1970	42,966	15,548	3,620	62,134
1971	43,167	16,383	3,640	63,190
1972	44,190	17,356	3,774	65,320
1973	43,429	17,152	3,587	64,168
1974	41,407	15,885	3,362	60,654
1975	38,210	15,773	3,138	57,121
1976	37,970	16,506	2,977	57,453
1977	40,095	16,551	2,813	59,459
1978	46,091	16,182	2,573	61,846
1979	52,598	16,036	2,815	71,449
1980	58,742	17,243	2,751	78,736

Source: Data for 1970-1979 from *Engineering Manpower Bulletin #50*, November 1979, Engineers Joint Council, New York, N.Y. 1980 data from Engineering Manpower Commission, New York, N.Y.

7. Engineering Manpower Commission, New York, 1980, and *New York Times Magazine*, June 28, 1981, p. 53.

8. *New York Times Magazine*, June 28, 1981 page 4.

9. *Engineering Education News*, vol. 8, no. 10, April 1982.

10. Stephen Kahne, "A Crisis in Electrical Engineering Manpower," *IEEE Spectrum*, July 1981 p. 50.

THE TOP TEN PRODUCERS OF ENGINEERS

Bachelor's		Master's		Doctor's	
Purdue U.	1442	Stanford U.	770	MIT	162
U. of Illinois	1257	MIT	673	U. Illinois	156
Penn. State	1136	Cal. Berkeley	571	Cal. Berkeley	131
Texas A&M	1081	Poly Inst. NY	512	Stanford U.	110
Georgia Tech	985	U. Michigan	415	Purdue U.	91
U. Michigan	917	U. Illinois	400	Cornell U.	82
U. Missouri, C.	794	U.S.C.	397	Ohio State	63
U. Missouri, R.	788	G.Washington U.	317	U.S.C.	62
U. Washington	765	Purdue U.	295	UCLA	58
Virginia Tech	762	Georgia Tech	294	Northwestern	53

Source: *Engineering Education*, April 1981, p. 713.

11. Interview, Cambridge, Massachusetts, August 1981.

12. "Survey Pegs Faculty Shortage at 10 Percent," *Engineering Education News*, January 1982.

13. AEA, op. cit., p. 1.

14. The top ten producers of engineers are as follows:

15. Interview, Palo Alto, California, September 1981.

16. Interview, spring 1982.

17. *Manpower Comments*, November 1981, p. 26.

18. Daniel C. Drucker, Dean College of Engineering, University of Illinois at Urbana-Champaign, before the Subcommittee on Science Research and Technology of the Committee on Science and Technology of the U.S. House of Representatives, March 3, 1981.

19. National Academy of Engineering, *Issues in Engineering Education*, Washington, D.C.: NAE, April 1980, p. 13.

20. National Science Foundation, "Federal Funding for Universities," quoted by Paul Gray, in a presentation to the Conference on Engineering Education, Massachusetts High Technology Council and the American Electronics Association, Natick, Massachusetts, February 10, 1982.

21. Data from National Assessment of Educational Progress, NCES-HEW. Quote from National Assessment of Educational Progress, NCES-HEW, as reported in *Manpower Comments*, September 1981, p 30.

22. "Bell Gives Some Failing Marks to American Education System," *The Boston Globe*, August 29, 1981.

23. Comments at National Governors' Association meeting, Task Force on Technological Innovation, Washington, D.C., February 1982.

24. *IEEE Spectrum*, November 198l, p. 69.

25. The Finniston Report, *Engineering Our Future*, Report of the Committee of Inquiry into the Engineering Profession, London, January, 1980.

26. Efthalia Walsh and John Walsh, "Crisis in the Science Classroom", *Science 80* 1, no. 6, pp. 17-22.

27. *Education USA*, May 4, 1981, p. 285.

MEN VS WOMEN: Intended Areas of Study

Predominantly Male	Percent **Male**	**Predominantly Female**	Percent **Female**
Military Science	91%	Home Economics	91%
Engineering	86%	Library Science	87%
Architecture		Foreign	
Environmental	77%	Languages	84%
Design	76%	Psychology	81%
Geography	76%	Education	79%

Forestry Conservation	74%	Art	73%
Physical Sciences	73%	Theater Arts	72%
Philosophy and Religion	65%	Health and Medical	71%
Agriculture	64%	English Literature	70%
History & Cultures	62%	Ethnic Studies	63%

Source: The College Board, 1981

28. National Science Foundation, *Science & Engineering Education in the 1980's & Beyond*, Washington, D.C., October 1980, p. 59.

29. Comments at National Governors' Association meeting, Task Force on Technological Innovation, Washington, D.C., February 1982.

30. Comments at National Governors' Association meeting, Task Force on Technological Innovation, February 1982, Washington, D.C.

31. The intended areas of study of men in contrast to women is given as follows:

32. Jewell Plummer Cobb, "Seven Filters Hampering Women in Science," *U.S. Woman Engineer*, February 1981, p. 1.

33. Figures from John C. Hoy and Melvin H. Bernstein, editors, *Business and Academia: Partners in New England's Economic Renewal*, University Press of New England, Hanover and London, 1981.

34. Cobb, "Seven Filters Hampering Women in Science," p. 2.

35. Stanford Women in Engineering, W.I.S.E.(Women in Science and Engineering), Stanford University, 1981.

36. Ibid.

37. Diane T. Rudnick and Susan E.D. Kirkpatrick, "Male and Female ET Students: A Comparison," in *Engineering Education*, May 1981, p. 765. For 1980-81, the figures were 6,557 women versus 56,378 men at the bachelor's level; 1,225 women versus 16,418 men at the master's, and 90 women versus 2,751 men at the doctoral level, according to the Engineering Manpower Commission of the American Association of Engineering Societies in "Engineering and Technology Degrees 1981."

CHAPTER 5

1. *$290 estimate:* *Electronics*, September 22, 1981
   $324 estimate: *Aviation Week & Space Technology*, September 21, 1981
   $500 estimate: Larry W. Sumney, Director VHSIC Program Office, Department of Defense, May 26, 1981: "At present funding levels, more than $500 million will be spent over five years."

2. J.R. Suttle, *Manpower Comments*, Washington, D.C., October 1981.

3. National Action/Research on the Military Industrial Complex (NARMIC), Philadel-

phia, September 1981. The role of DoD in semiconductor consumption is given in the following table:

U.S. PRODUCTION OF SEMICONDUCTORS for Defense Consumption

Year	Total[a]	Defense	Defense as a Percent of Total
1955	$ 40	$ 15	30%
1956	60	32	36
1957	151	54	36
1958	210	81	39
1959	396	180	45
1960	542	258	39
1961	565	222	39
1962	571	219	38
1963	594	196	33
1964	635	157	25
1965	805	190	24
1966	975	219	22
1967	879	205	23
1968	847	179	21
1969	1457	247	17
1970	1337	275	21
1971	1519	193	13
1972	1912	228	12
1973	3458	201	6
1974	3916	344	9
1975	3001	239	8
1976	4968	480	10
1977	4583	538	12

[a]Figures are in millions of dollars.

Source: "Innovation Competition and Government Policy in the Semiconductor Industry," Boston, Mass.: Charles River Associates, March, 1980.

4. *VHSIC: A New Era in Electronics*, Conference Proceedings, Boston, Mass., January 21-22, 1980.

5. Interview, fall 1981.

6. "Science and Engineering Manpower Forecasting: Its Use in Policymaking," Report by the U.S. General Accounting Office, Washington, D.C.: GAO, June 27, 1979.

7. The numbers on Defense Procurement and Research, Development, Testing, and Evaluation (RDT&E) outlays are as follows:

DEFENSE PROCUREMENT AND RESEARCH, DEVELOPMENT, TESTING, AND EVALUATION OUTLAY TRENDS (in millions of dollars)

	Procurement		Research, Development, Test & Evaluation	
FY	Outlays	Percentage change	Outlays	Percentage change
1965	$11,839	-22.9%	$ 6,236	-11.2%
1966	14,339	+21.1	$ 6,259	+ 0.4
1967	19,012	+32.6	$ 7,160	+14.4
1968	23,283	+22.5	$ 7,747	+ 8.2
1969	23,988	+ 3.0	$ 7,457	- 3.7
1970	21,584	-10.0	$ 7,166	- 3.9
1971	18,858	-12.6	$ 7,303	+ 1.9
1972	17,131	- 9.2	$ 7,881	+ 7.9
1973	15,654	- 8.6	$ 8,157	+ 3.5
1974	15,241	- 2.6	$ 8,582	+ 5.2
1975	16,042	+ 5.3	$ 8,866	+ 3.3
1976	15,964	- 0.5	$ 8,923	+ 0.6
1977	18,178	+13.9	$ 9,795	+ 9.8
1978	19,976	+ 9.9.	$10,508	+ 7.3
1979	25,404	+27.2	$11,152	+ 6.1
1980	29,021	+14.2	$13,127	+17.7
1981	35,191	+21.3	$15,278	+16.4
1982B	41,325	+17.4	$18,299	+19.8
1983B	55,144	+33.4	$22,200	+21.3
1984F	70,022	+27.0	$25,608	+15.4
1985F	88,262	+26.0	$29,343	+14.6

Note: FY through 1976 end in June, FY 1977 and after end in September. Data for the July-September 1976 transition are omitted for clarity, but are available.

B = Budgeted; F = Long-range Forecast.

Source: *The Rosen Electronics Letter*, February 22, 1982.

8. *The Rosen Electronics Letter*, February 22, 1982.

9. Engineering Manpower Concerns, Hearings before the Committee on Science and Technology, U.S. House of Representatives, 97th Congress, Washington, D.C., October 6-7, 1981.

10. *Fortune*, November 2, 1981, p. 112.

11. "Criticism Rises on Reagan's Plan for 5-year Growth of the Military", *New York Times*, March 22, 1982, p. 1.

12. David L. Blond, "On the Adequacy and Inherent Strengths of the United States Industrial and Technological Base: Guns versus Butter in Today's Economy", Washington, D.C., Office of the Secretary of Defense, Program Analysis and Evaluation, Department of Defense, 1980.

13. *Business Week*, June 8, 1981, p. 112.

14. Memorandum to the Secretary of Defense from Russell Murray 2nd, Assistant Secretary of Defense, Program Analysis and Evaluation, November 24, 1980.

15. Ibid.

CHAPTER 6

1. National Governors' Assocation, Task Force on Technological Innovation, January 29, 1982.

2. Other reasons are according to Erich Bloch in personal correspondence, "the increasing cost of research, the need for a viable university system, the requirements for an increasing number of technically educated professionals, and world competition – especially from Japan."

3. Semiconductor Research Cooperative Working Proposal, draft manuscript, November 1981.

4. He has since moved to a similar setting in a new headquarters building.

5. Personal interview, spring 1982.

6. Personal interview, spring 1982.

7. Personal interview, winter 1982.

8. The six participating institutions are the University of North Carolina at Chapel Hill, Duke University, University of North Carolina at Charlotte, North Carolina A & T State University, North Carolina State University, Research Triangle Institute.

9. Personal interview, May 1982.

10. The eight Massachusetts educational institutions were Boston University, MIT, Northeastern University, Southeastern Massachusetts University, Tufts, University of Lowell, University of Massachusetts, and Worcester Polytechnic. Among initial industry discussions were representatives from Alpha Industries, Analog Devices, Digital Equipment Corporation, the local branch of Hewlett-Packard, M/A Com, Raytheon, Sperry, Sprague Electric, Unitrode, and Wang Laboratories.

11. Personal interview, May 1982.

12. Personal interview, May 1982.

13. In refering to IBM and the other American computer companies during an interview with the authors, Professor Moses coined the phrase "IBM and BUNCH": IBM and Burroughs, Univac, National Cash Register, Control Data Corporation, and Honeywell. "What can I do with DEC and Apple?", he mused.

14. Personal interview, May 1982.

15. Personal interview, May 1982.

CHAPTER 7

1. In addresses presented to the Conference on Engineering Education, Massachusetts High Technology Council, and the American Electronics Association, Natick, Massachusetts, February 10, 1982.

2. All figures are from the *Digest of Educational Statistics for 1981*, Washington, D.C.: National Center for Educational Statistics.

3. Ibid.

Institution	Total R&D (in millions of dollars)	Federal Share	Federal percent of total
MIT	141,596	120,971	85%
University of Wisconsin-Madison	122,239	78,096	84
University of California-San Diego	107,750	96,375	89
University of Michigan	107,035	67,396	63
University of Minnesota	106,077	61,150	58
Stanford University	101,681	91,511	90
Cornell University	100,295	67,001	67
University of Washington	98,967	83,276	84
Harvard University	89,008	67,374	76
Columbia University	82,831	67,012	81
University of Pennsylvania	81,961	58,941	72
University of Illinois-Urbana	75,972	44,670	59
Johns Hopkins University	75,592	64,732	86
University of California-L.A.	75,496	60,102	80
University of California-Berkeley	75,344	53,520	71
University of Texas-Austin	69,695	45,661	66
Pennsylvania State University	64,054	40,859	64
Texas A&M University	63,271	28,127	44
Ohio State University	62,747	35,181	56
Michigan State University	62,666	27,827	44
TOTAL Leading 20	1,764,277	1,259,773	71
TOTAL Leading 100	4,291,358	2,860,418	67
TOTAL All Institutions	5,176,930	3,428,957	66

4. The totals are: public and private = $47.0 billion for current fund revenue; if "plant fund revenue" are added in, total = $ 51.8 billion. In 1980-81, the total support to public = $ 43.9, private = $ 21.1. The total is $65 billion. The federal share was down one percentage point – 12 percent of public and 18 percent of private.

5. J. Krakower, "Federal Support of R&D at Leading Research Universities," *The NCHEMS Newsletter*, Summer, 1981 The top twenty leading research universities for 1979 were as follows:

6. All numbers from J. Krakower, "Federal Support of R&D at Leading Research Universities," *The NCHEMS Newsletter*, Summer, 1981.

7. *Boston Globe*, January 18, 1982; and *New York Times* January 20, 1982, and February 16, 1982. According to the Office of the Management and Budget, estimated federal R&D funding for 1981 and 1982 by department are as follows:

ESTIMATED FEDERAL R&D FUNDING (1981 and 1982): (in millions of dollars)

	1980[a]	1981	1982	Carter 1982
Defense	16,494	20,553	24,469	
NASA	5,407	5,841	6,513	
Commerce	5,276	4,793	4,157	
HHS	3,973	3,972	4,122	
NSF	964	961	1,033	vs 1,353.5
USDA	773	807	883	
Interior	424	397	371	
Transportation	420	329	366	
EPA	326	317	230	
NRC	227	223	220	
AID	156	160	186	
VA	147	137	145	
Education	91	74	76	
Other	354	279	272	
	35,032	38,843	42,998	

Source: Office of Management & Budget
[a]Actual figures. All years are fiscal years.

8. Cited by David Knapp, President of the University of Massachusetts (Amherst) in an address presented to the Conference on Engineering Education, Massachusetts High Technology Council and the American Electronics Association, Natick, Massachusetts, February 10, 1982.

9. State activities to encourage technological innovation were listed in a national governors' conference report as follows:

STATE ACTIVITIES TO ENCOURAGE TECHNOLOGICAL INNOVATION (1981) (in millions of dollars)

State	TOTAL	State	TOTAL
Arkansas	$ 239	Michigan	$ 1,200
Arizona	1,917	Minnesota	35
California	27,350	Missouri	40
Colorado	950	Nebraska	200
Connecticut	12,000	New Jersey	200
Georgia	350	New Mexico	2,000
Hawaii	50	New York	2,545
Illinois	28,129	No. Carolina	29,900
Indiana	272	Pennsylvania	1,255
Kentucky	621	Tennessee	572
Maryland	806	Texas	1,825
Massachusetts	1,190	Virginia	10
		TOTAL (1981)	$113,656

Three states account for 75 percent or $85,389

Source: National Governors' Conference, Washington, D.C., February 1982.

10. All figures are from the *Digest of Educational Statistics for 1981*, Washington, D.C.: National Center for Educational Statistics.

11. Edmund C. Cranch, in an address presented to the Conference on Engineering Education, Massachusetts High Technology Council and the American Electronics Association, Natick, Massachusetts, February 10, 1982.

12. President Kenneth G. Ryder, in an address presented to the Conference on Engineering Education, Massachusetts High Technology Council and the American Electronics Association, Natick, Massachusetts, February 10, 1982,

13. For example, by John Silber, President of Boston University

14. The following table shows the breakdown of corporate giving by recipient:

WHERE THE DOLLAR WENT:
A Survey of 732 Companies (1980)

Education[a]	$375,877	37.8%
Health & Welfare	337,866	34.0
Civic Activities	116,788	11.7
Culture & Art	108,673	10.9
Other	55,452	5.6
TOTAL	$994,626	100.0%

[a]Higher Education 24.4% of total

Source: Department of Commerce, Internal Revenue Service, Council for Financial Aid to Education, October 1981.

15. Council for Financial Aid to Education, New York, October 1981, and "Gifts to Charity Rise 12.3% to $53.6 billion, a Record," *The New York Times, April 9, 1982.*

16. The calculation would be as follows: Since 6 percent of students are in engineering, the prorata cost of engineering education would be 6 percent of $65 billion or $3.9 billion. However, the cost of educating an engineering student is considerably above the average per student cost. The AEA figure for support to engineering education is calculated as follows: $200 billion in electronics sales times 10 percent R&D times the 2 percent proposed by AEA = $400 million.

17. Interview, winter 1982.

18. "Private Industry on Campus," *New York Times*, November 5, 1981.

19. David Sanger, "Harvard Shuns the Apple but Doesn't Step on the Serpent," *New York Times*, February 21, 1982.

20. Cost estimates are taken from Engineering Education & Accreditation Committee of the Engineer's Council for Professional Development as published in *Engineering Education*, February 1980. The estimates were gathered from a survey of 41 institutions nationwide with a total of 143 specific engineering programs. The dollar costs are for the academic year 1979-1980.

21. For example, an estimate of the salary differential, made by the MIT President's Office, is shown below:

FALLING BEHIND IN A GAME OF CATCH-UP

Year	Weighted Average Salary Assistant Professor at thirty-six Institutions (11 Months)	Engineering Ph.D. (MIT) Average Starting Salary (Industry)
1978	$19,700	$22,000
1979	21,200	24,900
1980	22,500	26,200
1981	24,700	30,000
1982	n.a.	33,500

Source: MIT President's Office

22. Daniel Drucker, Dean of Engineering at the University of Illinois, Statement before the Subcommittee on Science Research and Technology of the Committee on Science and Technology of the U.S. House of Representatives relative to NSF Authorization for FY82, Washington, D.C., March 3, 1981.

23. If industry employment requirements increased by 15 percent per year, this implies a doubling of manpower requirements in a decade. The American Electronics Association's study shows the need to triple output within five years.

24. Peter J. Denning, ed., *Communications of the ACM*, June 1981, p. 374.

25. Cost estimates are taken from Engineering Education & Accreditation Committee of the Engineer's Council for Professional Development as published in *Engineering Education*, February 1980. The estimates were gathered from a survey of 41 institutions nation-

wide with a total of 143 specific engineering programs. The dollar costs are for the academic year 1979-1980.

26. Edgar Faure et al., *Learning To Be*, Paris: UNESCO, 1972.

27. *The ICIS Guide to Educational Technology*, The Learning Project and the International Center for Integrative Studies, 45 West 18th Street, New York, N.Y. 10011, 1981.

28. National Academy of Engineering, "Educational Technology in Engineering," Washington D.C.: National Academy Press, 1981.

29. *Monitor* Vol 5, no.1, January 1982. Published by the AMCEE.

30. Ibid.

31. John A. Curtis, "Instructional Television Fixed Service: A Most Valuable Educational Resource," in *Educational Telecommunications Delivery Systems*, John A. Curtis and John M. Biedenbach, eds, Washington, D.C.: American Society of Engineering Education, 1980.

32.Thomas E. Shanks and John L. Hochheimer, "Planning Instructional Television for the Continuing Education of Engineers: A Selected Review and Application of the Literature since 1967," paper presented to the International Communication Association — 30th Annual Convention, Acapulco, Mexico, May 20, 1980.

33. Interview, fall 1981.

CHAPTER 8

1. Fred Hechinger, *New York Times*, January 26, 1982.

2. Interview, winter 1981.

3. See, for example, James Botkin, Mahdi Elmandjra, and Mircea Malitza, *No Limits to Learning: A Report to The Club of Rome*, Oxford: Pergamon Press Ltd., 1979.

4. *The New Liberal Arts — An Exchange of Views*, New York: Alfred P. Sloan Foundation, August, 1981, p. 33.

5. C.P. Snow, *The Two Cultures: And a Second Look*, Cambridge: Cambridge University Press, 1964.

6. Gerald Holton, "Where Is Science Taking Us?", 1981 Jefferson Lectures, National Endowment for the Humanities (draft, to be published by Cambridge University Press, 1983). Italics added.

7. Fritz Machlup, *Knowledge: Its Creation, Distribution, and Economic Significance*, vol. I: *Knowledge and Knowledge Production*, Princeton, N.J.: Princeton University Press, 1980.

On the question of subjecting science to social goals, Fritz Machlup in his book, *Knowledge and Knowledge Production*, expresses his doubts that the humanities are capable of "humanizing" technology. He dismisses such notions as no more than a "sales pitch" used to justify periodic government support for the humanities (page 83). But if this is just a sales pitch, it has quite a number of buyers already.

8. John C. Hoy and Melvin H. Bernstein, editors, *Business and Academia: Partners in New England's Economic Renewal*, Hanover and London: University Press of New England, 1981, pp 130-131.

9. Gerald Holton, "Where Is Science Taking Us?," 1981 Jefferson Lectures, National Endowment for the Humanities (Cambridge: Cambridge University Press, forthcoming, 1983).

10. *The New Liberal Arts – An Exchange of Views*, New York: Alfred P. Sloan Foundation, August 1981, p. 15.

11. J. Paul Hartman and Robert D. Kersten, "Engineering Course Work for Non-Engineers: A Decade of Experience", University of Central Florida (formerly Florida Technological University), in *Engineering Education*, December 1980.

CHAPTER 9

1. S. Alexander Rippa, *Education in a Free Society*, New York: David McKay Company, Inc., 1971.

2. Dean Wilson Kuykendall, *The Land-Grant College – A Study in Transition*, Ph.D. dissertation, Graduate School of Education, Harvard University, 1946, p. 80.

3. Alexis de Tocqueville, *Democracy in America* (1826), cited in Kuykendall, op. cit., p. 22.

4. Kuykendall, op. cit., p. 13.

5. Kuykendall, op. cit., p. 78.

6. Report prepared by the staff of Governor James Hunt, Jr., *Federal-State Science Policy: Present and Future*, Background Paper, National Governors' Association, Task Force on Technological Innovation, Washington D.C., February 1982.

Other Sources on the Morrill Act

Fred J. Kelly, "Land-Grant Colleges and Universities – A Federal-State Partnership," Bulletin 1952, no. 21, Federal Security Agency, Washington, D.C.: Office of Education, 1952

S. Alexander Rippa, *Education in a Free Society: An American History*, University of Vermont, New York: David McKay Company Inc. 1971, p. 310.

"The Development of the Land-Grant Colleges and Universities and Their Influence on the Economic and Social life of the People," Addresses Commemorating the Centennial of the First Morrill Act, *West Virginia University Bulletin*, Morgantown, West Virginia, 1962.

CHAPTER 10

Governor James B. Hunt, Jr.

1. See U.S. Congress, House Committee on Science and Astronautics, Subcommittee on Science, Research and Development, report entitled *Toward a Science Policy for the United States*, Washington, D.C.: U.S. Government Printing Office, 1970, p. 81.

2. Willis H. Shapley, Albert H. Teich and Gail J. Breslow, *Research and Development, AAAS Report VI*, Washington, D.C.: American Association for the Advancement of Science, 1981, pp. 17, 87.

3. Izak Wirzup, University of Chicago, in a letter report to the National Science Foundation in 1979.

4. Willis H. Shapley, Albert H. Teich, Gail J. Breslow and Charles B. Kidd, *Research and Development, AAAS Report V*, Washington, D.C.: American Association for the Advancement of Science, 1980, pp. 88-90.

5. Louis Harris, The Crisis of '82, Presentation at the Yale Political Union, Yale University, New Haven, Connecticut, September 10, 1981.

6. See *State Activities to Encourage Technological Innovation*, a report prepared for the National Governor's Association by the State of California, August 1981. Also, a record of progress is maintained by the "State Science, Engineering and Technology Program" National Science Foundation, Washington.

7. Robert E. Evenson, Paul E. Waggoner and Vernon W. Ruttan, "Economic Benefits from Research: An Example from Agriculture," *Science*, Washington, D.C.: AAAS, vol. 205, September 1979, p. 1103.

8. A.N. Whitehead, *Science and the Modern World*, New York: Macmillan, 1925, p. 199.

Professor Robert M. Hexter

1. See *C & E News*, July 27, 1981, p. 63, which shows that R & D spending has been decreasing as a percent of net sales for all U.S. industry over the period 1966 to 1981.

2. The Shell Development Laboratories were located in Emeryville, California.

INDEX

ABOUT THE AUTHORS

JAMES W. BOTKIN. Internationally acclaimed author of *No Limits To Learning: A Report to The Club of Rome.* Doctorate in computer based systems from the Harvard Business School. Consultant to industry, government, and universities on technology, management, and learning. Cofounder of the Forum Humanum, an international network on futures research. Former academic director of the Salzburg Seminar.

DAN DIMANCESCU. International consultant and writer on high technology strategy and policy issues. Director of U.S., Japanese, and European urban planning and management studies. Editor of *Rites of Way*, an analysis of changing transportation priorities in the U.S. Educated at Dartmouth College, the Fletcher School of Law and Diplomacy, and the Harvard School of Business Administration.

RAY STATA. Founder and president of Analog Devices, Inc. in Norwood, Massachusetts. In 1979 initiated the founding of the Massachusetts High Technology Council; currently Chairman of its Human Resources Committee. Currently serving as a member of the Massachusetts Board of Regents for Higher Education. Educated at the Massachusetts Institute of Technology in electrical engineering.

JOHN McCLELLAN. Recent consulting projects include a review of educational technologies for the U.S Department of Education. A former software consultant and teacher at elementary and university levels. Completing doctoral work on collective learning in the Future Studies Program at the University of Massachusetts. Finalist in the 1979 Mitchell Prize Contest.

ACKNOWLEDGMENTS

The editorial coordination for this book was by the Technology and Strategy Group, a consulting and writing team of which James Botkin and Dan Dimancescu are principals. The authors wish to thank Dr. An Wang and Wang Laboratories for the generous provision of a word processor on which the entire manuscript was prepared and which immeasurably aided the efficiency of the coauthorship.

The project was administered by the International Center for Integrative Studies (ICIS), a nonprofit service, research, and educational center in New York City (45 West 18th Street, New York, N.Y. 10011). Special thanks are due to ICIS Vice-President George Christie for his valued suggestions and to the Director of the ICIS Information Center, Helen Robinette, without whose dedication in tracking down obtuse sources of information, this book would not have been possible.

And were it not for the initial enthusiasm and encouragement of Carol Franco, editor of Ballinger Publishing Company, many months of research and writing might never have come to fruition.

Comments from readers are invited, and can be sent to the coauthors at:

Technology and Strategy Group
Harvard Square
50 Church Street
Cambridge, Massachusetts 02138

Jacket and chart design by Turnbull & Company, Cambridge, Massachusetts.

The text was typeset in Janson, a typeface based on the original cut by Nicholas Kis in 1690. Chapter heads are typeset in Granjon, designed by George W. Jones for Merganthaler in honor of French type designer Robert Granjon. It is based on the original cut in Frankfort, Germany in 1592.

Our thanks to Fred Alexander of Alexander Typesetting, Inc., Indianapolis, Indiana, for his advice and patience in laboring through a difficult manuscript.

ACKNOWLEDGMENTS